Lecture Notes in Mathematics

Edited by A. Dold and B. Eckmann

724

David Griffeath

Additive and Cancellative Interacting Particle Systems

Springer-Verlag
Berlin Heidelberg New York 1979

Author

David Griffeath
Dept. of Mathematics
University of Wisconsin
Madison, WI 53706
USA

AMS Subject Classifications (1970): 60K35

ISBN 3-540-09508-X Springer-Verlag Berlin Heidelberg New York
ISBN 0-387-09508-X Springer-Verlag New York Heidelberg Berlin

Printing and binding: Beltz Offsetdruck, Hemsbach/Bergstr.
2141/3140-543210

Preface

These notes are based on a course given at the University of Wisconsin in the spring of 1978. The subject is (stochastic) interacting particle systems, or more precisely, certain continuous time Markov processes with state space $S = \{\text{all subsets of } Z^d\}$. This area of probability theory has been quite active over the past ten years: a list of references, by no means comprehensive, may be found at the end of the exposition. In particular, several surveys on related material are already available, among them Spitzer (1971), Dawson (1974b), Spitzer (1974b), Sullivan (1975), Georgii (1976), Liggett (1977) and Stroock (1978). There is rather little overlap between the present treatment and the above articles, and where overlap occurs our approach is somewhat different in spirit.

Specifically, these notes are based on graphical representations of particle systems, an approach due to Harris (1978). The basic idea is to give explicit constructions of the processes under consideration with the aid of percolation substructures. While limited in applicability to those systems which admit such representations, Harris' technique manages to handle a large number of interesting models. When it does apply, the graphical approach has several advantages over alternative methods. First, since the systems are constructed from "exponential alarm clocks," the existence problem does not arise. Also, the uniqueness problem can be handled with much less difficulty than for more general particle systems. Another appealing feature is the geometric nature of the representation, which leads to "visual" probabilistic proofs of many results. Finally, there is the matter of coupling. One of the basic strategies in studying particle systems is to put two or more processes on a joint probability space for comparison purposes. Graphical representations have the property that processes starting from arbitrary initial configurations are all defined on the same probability space, in such a way that natural couplings are often embedded in the construction. This is a major conceptual simplification in many arguments. Altogether, Harris' approach makes the material easily accessible to a gifted graduate student having a familiarity with the elementary theory of Markov chains and processes.

The development is divided into four chapters. Chapter I contains basic notation, general concepts and a discussion of the major problems in the field of interacting particle systems. It also includes a description of the percolation substructures which are used to define the processes we intend to study. Chapter II is devoted to additive systems. The "lineal" additive systems were introduced by Harris (1978). We also consider "extralineal" additive systems. General ergodic and pointwise ergodic theorems are proved. Among the specific models treated in some detail are contact processes, voter models and coalescing random walks. Chapter III deals with cancellative systems, a second large class of models which admit graphical representation. There are analogous general ergodic theorems for this class. Specific topics include an application to the stochastic Ising model, and limit theorems for generalized voter models and annihilating random walks. In Chapter IV we discuss the uniqueness problem for additive and cancellative systems We have closen to present this material last, since uniqueness questions seem rather esoteric in comparison with the important problems of ergodic theory. The graphical approach shows how nonuniqueness can arise when there is "influence from ∞."

A great deal of the material in these notes has appeared in recent research papers by many authors. At the end of each section is a paragraph entitled "Notes" which identifies the sources of the results contained therein. All references are to the Bibliography which follows Chapter IV.

I would like to acknowledge my gratitude to many mathematicians for their contributions, especially M. Bramson, D. Dawson, Sheldon Goldstein, L. Gray, T. Harris, R. Holley, H. Kesten, T. Liggett, F. Spitzer and D. Stroock. Let me also thank the various Soviet mathematicians whose pioneering work on closely related discrete time systems was a major source of inspiration for the continuous time theory. A sampling of their publications is included in the Bibliography. Finally, my thanks go out to Richard Arratia, Steve Goldstein and Arnold Neidhardt for their many comments and corrections as these notes were taking shape.

David Griffeath
Madison, Wisconsin
August, 1978

CONTENTS

CHAPTER I: INTRODUCTION

1. Preliminaries.

Throughout the exposition we will use the following notation:

Z^d = the d-dimensional integer lattice $(d \geq 1)$;

$x, y, z \in Z^d$ are called sites.

$S = \{\text{all subsets of } Z^d\}$,

$S_0 = \{\text{all finite subsets of } Z^d\}$,

$S_\infty = \{\text{all infinite subsets of } Z^d\}$.

$A, B, C \in S$ are called configurations. Write

$$A(x) = 1 \quad \text{if} \quad x \in A,$$
$$= 0 \quad \text{if} \quad x \notin A.$$

Λ will always be a finite configuration, i.e. $\Lambda \in S_0$;

$|\Lambda|$ is the cardinality of Λ.

Important finite configurations are the n-box $b_n(x)$ centered at $x \in Z^d$:

$$b_n(x) = \{y = (y_1, \cdots, y_d) : |y_\ell - x_\ell| \leq n \text{ for } 1 \leq \ell \leq d\},$$

and the block $[x, y] \subset Z$, $x, y \in Z$:

$$[x, y] = \{z : x \leq z \leq y\}.$$

One useful abuse of notation is to write x instead of $\{x\}$ for the singleton configuration at site x; we will do this whenever it is convenient.

$T = [0, \infty)$ is the (continuous) time parameter set;

$r, s, t, u \in T$ are times.

Our objects of study will be certain continuous time S-valued Markov processes, or particle processes. Such a process will be written as

$$(\xi_t^A)_{t \in T}, \quad \text{or simply} \quad (\xi_t^A).$$

Here ξ_t^A is the configuration of the process at time t, and A is the initial state, i.e. $\xi_0^A = A$. We say that there is a particle at site x at time t if $x \in \xi_t^A$.

Other notations for particle processes are (η_t^A) and (ζ_t^A). A family $\{(\xi_t^A); A \in S\}$ of particle processes will be called a <u>particle system</u>. P and E will be the probability law and expectation operator respectively governing such a system.

Some additional notation:

$\mathfrak{m} = \{$all probability measures on $S\}$;

$\mu, \nu \in \mathfrak{m}$ are often called <u>distributions</u>.

$\delta_A \in \mathfrak{m}$ is the measure concentrated at $A \in S$. For $0 \le \theta \le 1$, $\mu_\theta \in \mathfrak{m}$ is the Bernoulli product measure such that $\mu_\theta(\{A : x \in A\}) = \theta$ for all x. Thus $\mu_0 = \delta_\emptyset$, $\mu_1 = \delta_{Z^d}$. Any $\mu \in \mathfrak{m}$ is uniquely determined by its restriction to cylinder sets $\{A : A \cap \Lambda = \Lambda_0\}$, $\Lambda_0 \subset \Lambda \in S_0$. In fact, by inclusion-exclusion, μ is uniquely determined by its <u>zero function</u> $\varphi^\mu : S_0 \to [0,1]$,

$$\varphi^\mu(\Lambda) = \mu(\{A : A \cap \Lambda = \emptyset\}) .$$

A particle process started from an initial distribution μ will typically be written as (ξ_t^μ); thus $\xi_t^A = \xi_t^{\delta_A}$. For each $t \ge 0$, the distribution $\mu P^t \in \mathfrak{m}$ of ξ_t^μ is given by

$$\mu P^t(\cdot) = P(\xi_t^\mu \in \cdot),$$

with zero function

$$\varphi_t^\mu = \varphi^{\mu P^t} .$$

The transition mechanisms for the Markov systems we propose to study may be prescribed in terms of <u>jump rates</u>

$$c = \{c(A, A') : A \Delta A' \in S_0\} .$$

Intuitively, $c(A, A')$ represents an exponential rate at which the configuration of particles "tries to jump" to A' from A; we require $A \Delta A' \in S_0$ to ensure that the configuration of the infinite system changes at only finitely many sites at any one time. The precise formulation of jump rates will be deferred until Chapter 4. In the important special case where changes in configuration can only occur at one site at a time, $\{(\xi_t^A)\}$ is called a <u>spin system</u>. Such a system is described by its flip rates

$$c_x(A) = c(A, A \Delta x) ;$$

roughly,

$$c_x(A)dt \sim P(\xi^A_{dt}(x) \neq A(x)) .$$

We now introduce three simple particle systems: the basic one-dimensional coalescing and annihilating random walks and contact processes.

(1.1) Example. Basic coalescing random walks. Particles, one starting from each site of A, execute continuous time simple random walks on Z subject to an interference mechanism: whenever a particle jumps to a site which is already occupied, the two particles at that site coalesce into one. By a continuous time simple random walk we mean a motion which waits a mean-1 exponential time at each position, and then jumps one unit to the left or right, each with probability $1/2$. The particle motions in the coalescing random walks are independent except for the interference mechanism. This system may be constructed with the aid of a graphical representation as follows. Start with the space-time diagram $Z \times T$. For each site $x \in Z$ draw an infinite sequence of arrows with δ's on the tail: $\underset{x}{\delta}\!\!\longrightarrow\!\!\underset{x+1}{}$, from $(x, \tau^1_{1,x})$ to $(x+1, \tau^1_{1,x})$, $(x, \tau^2_{1,x})$ to $(x+1, \tau^2_{1,x})$, etc. The values $\tau^1_{1,x}$, $\tau^2_{1,x} - \tau^1_{1,x}, \cdots$ are taken to be independent exponential variables with mean $\frac{1}{2}$. Similarly put arrows with δ's on the tail: $\underset{x-1}{}\!\!\longleftarrow\!\!\underset{x}{\delta}$ from $(x, \tau^1_{2,x})$ to $(x-1, \tau^1_{2,x})$, $(x, \tau^2_{2,x})$ to $(x-1, \tau^2_{2,x})$, etc. for each $x \in Z$, where the $\tau^n_{2,x}$ occur at rate 1. A generic realization is shown in figure 1. Say there is a path up from (y, s) to (x, t) if there is a chain of upward vertical and directed horizontal edges in the resulting diagram which leads from (y, s) to (x, t) without passing (vertically) through a δ. The δ's may be thought of as obstructions to the flow (or "percolation") of liquid. Now define

(1.2) $$\xi^A_t = \{x : \text{there is a path up from } (y, 0) \text{ to } (x, t) \text{ for some } y \in A\} .$$

A little thought reveals that (ξ^A_t) is the basic coalescing random walks starting from A.

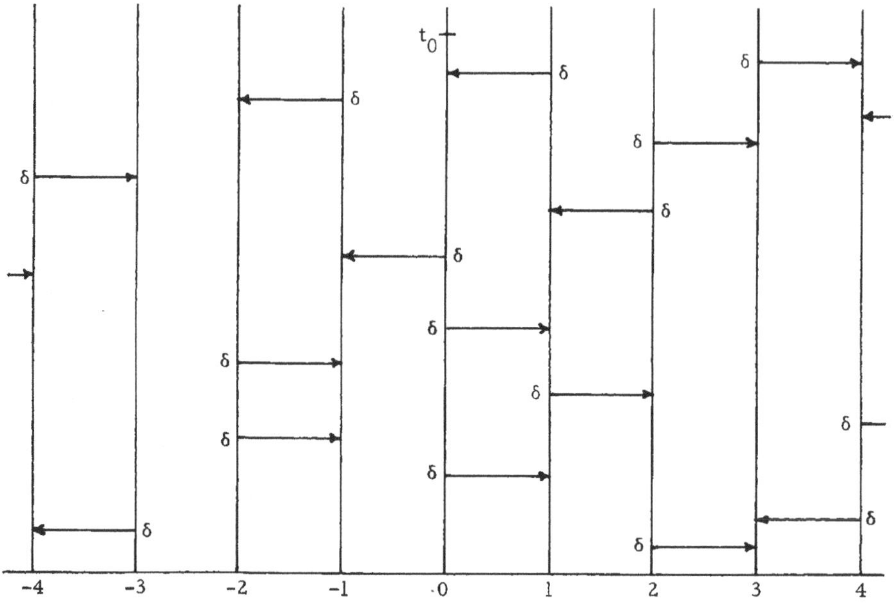

figure i .

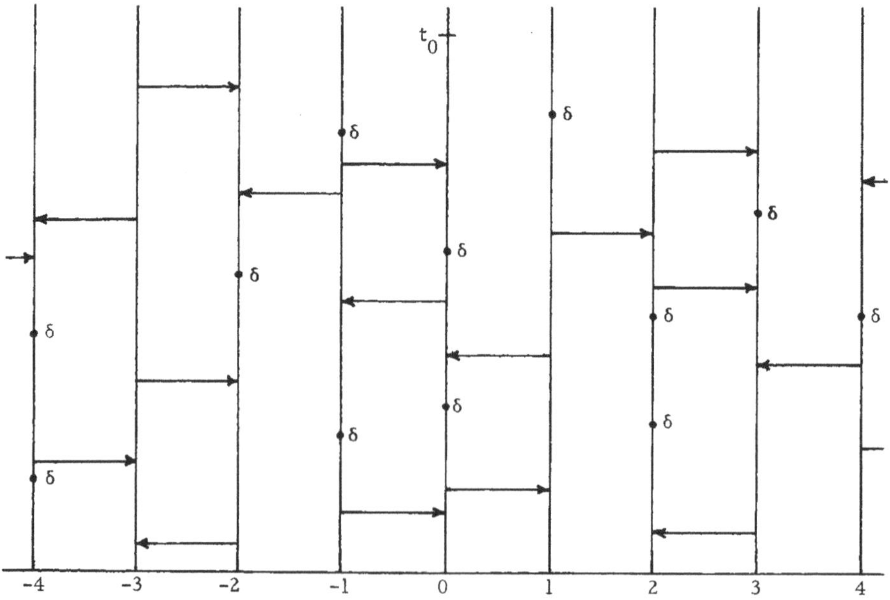

figure ii .

(1.3) Example. Basic annihilating random walks. This is the same as the first model, except that when a particle jumps to a site which is already occupied the two particles at that site annihilate one another. This second system can be defined using the same random graph, or percolation substructure, as in (1.1). If instead of (1.2) we take

(1.3) $$\eta_t^A = \{x : \text{the number of paths up to } (x,t) \text{ from } (A,0) \text{ is odd}\},$$

then (η_t^A) is the basic annihilating random walks starting from A.

(1.4) Problems. For the realization of the percolation substructure shown in figure i, and with $\{(\xi_t^A)\}$ and $\{(\eta_t^A)\}$ defined by (1.2) and (1.3) respectively, is $0 \in \xi_{t_0}^Z$? Is $0 \in \eta_{t_0}^Z$? It is obvious from the intuitive descriptions that $\emptyset$ is a trap (= absorbing state) for $\{(\xi_t^A)\}$ and $\{(\eta_t^A)\}$. Explain how this follows from (1.2) and (1.3). How many changes of the infinite configuration take place in (ξ_t^Z) and (η_t^Z) between $t = 0$ and $t = \varepsilon > 0$?

(1.5) Example. Basic contact systems. In this case $\{(\xi_t^A)\}$ is a spin system with flip rates

$$\begin{aligned} c_x(A) &= 1 && x \in A \\ &= \lambda\,|A \cap \{x-1, x+1\}| && x \notin A. \end{aligned}$$

$\lambda > 0$ is a parameter. We may think of the site x as infected when $x \in \xi_t^A$, and healthy when $x \notin \xi_t^A$. Thus infected sites recover at constant rate 1, while healthy sites are infected at a rate proportional to the number of infected neighbors. Thus the parameter λ is an infection index. Contact processes also admit graphical representations. Now three types of graphical device are attached to each $\{x\} \times T$. First, a sequence of δ's is put down at rate 1 (i.e. with independent exponential mean-1 times between successive δ's). These will have the effect of killing infection if it is present. Next, a sequence of directed arrows: $\underset{x-1 \quad x}{\longrightarrow}$ is put down at rate λ, and finally a sequence of arrows $\underset{x \quad x+1}{\longleftarrow}$ is also put down at rate λ. The resulting percolation substructure will look something like figure ii. Defining ξ_t^A by (1.2), in terms of this second substructure, we obtain the basic contact system.

(1.6) Problem. Let (ξ_t^0) be the basic contact process starting with only the origin infected. Show that for all sufficiently small positive λ, the infection dies out with probability one.

These notes will be devoted exclusively to particle systems which can be constructed from exponential random variables with the aid of percolation substructures. In this way we circumvent the first major problem in the theory of random interacting particle systems:

I. Existence: When is there a system $\{(\xi_t^A)\}$ with given jump rates c ?

A great many systems do not admit graphical representations in terms of percolation substructures, and for these the existence problem is nontrivial. A second fundamental question is:

II. Uniqueness: When is there a unique particle system $\{(\xi_t^A)\}$ with given jump rates?

Even for the models we will study, a precise formulation and treatment of this problem requires technical machinery; we therefore defer uniqueness questions until Chapter 4.

Once the system is well-defined, interest centers on:

III. Ergodic theory: What is the asymptotic behavior of the processes (ξ_t^A) as $t \to \infty$?

We now discuss the broad outlines of problem III. A familiar property of Markov processes is their "loss of memory" under appropriate assumptions on the transition mechanism. Starting from measure μ, it is common for μP^t to converge to an equilibrium, or invariant measure ν as $t \to \infty$. For particle systems the appropriate notion is that of weak convergence: $\lim_{t\to\infty} \mu_t = \nu$ $(t \in T$ or $t = 0,1,\cdots)$ if

$$\lim_{t\to\infty} \mu_t(\{A : A \cap \Lambda = \Lambda_0\}) = \nu(\{A : A \cap \Lambda = \Lambda_0\}) \quad \forall \Lambda_0 \subset \Lambda \in S_0 .$$

By inclusion-exclusion, this last is equivalent to:

$$\lim_{t\to\infty} \varphi^{\mu_t}(\Lambda) = \varphi^{\nu}(\Lambda) \qquad \forall \Lambda \in S_0 .$$

Say that ν is <u>invariant</u> for the system $\{(\xi_t^A)\}$ if $\nu P^t = \nu$ for each $t \in T$. The particle systems we study will almost always be <u>Feller</u>, in the sense that

$$\mu_\alpha P^t \to \mu P^t \quad \text{as} \quad \mu_\alpha \to \mu \quad \text{for each fixed } t \in T .$$

Any Feller system has at least one equilibrium. To see this, define the Cesaro measures 7

$$\mu C^t = \frac{1}{t} \int_0^t \mu P^s \, ds \qquad \mu \in \mathfrak{M}, \; t \in T .$$

Since S is compact (in the discrete product topology), so is $\mathfrak{M}$. Choose ν such that $\mu C^{t'} \to \nu$ as $t' \to \infty$, for some subsequence t'. Using the Feller property, it is easy to check that ν is invariant. Let $\mathcal{J}$ be the set of all invariant measures for $\{(\xi_t^A)\}$. We have seen that $\mathcal{J} \neq \emptyset$ in the case of a Feller system. The system is called <u>ergodic</u> if $\mathcal{J} = \{\nu\}$ for some $\nu \in \mathfrak{M}$, i.e. if it has a unique equilibrium. This is equivalent to

(1.7) $$\exists \nu \in \mathfrak{M} : \lim_{t\to\infty} \mu C^t = \nu \qquad \forall \mu \in \mathfrak{M} ..$$

Say that $\{(\xi_t^A)\}$ is <u>strongly ergodic</u> if

(1.8) $$\exists \nu \in \mathfrak{M} : \lim_{t\to\infty} \mu P^t = \nu \qquad \forall \mu \in \mathfrak{M} .$$

Clearly strong ergodicity implies ergodicity. When proving ergodic theorems we will invariably derive (1.8) rather than (1.7) in these notes. However, there is no known example of a particle system which satisfies (1.7) but not (1.8). For convenience we will usually omit the word "strong" in the statement of ergodicity results.

(1.9) <u>Problems</u>. Prove that (1.7) is equivalent to ergodicity. Show that (1.8) need only be checked for delta measures $\mu = \delta_A$, $A \in S$, to ensure strong ergodicity. Find a Feller family $\{(X_t^x);\; x \in S\}$ on a compact state space S, and an equilibrium ν, such that (1.7) holds but (1.8) does not.

(1.10) Problem. Let $\{(X_t^x)\}$ be a Feller family on a compact state space S, ν an invariant measure for the family. Show that the stationary process (X_t^ν) is Birkhoff ergodic if there is a set of states $S_\nu \subset S$ such that $\nu(S_\nu) = 1$ and

$$\lim_{t\to\infty} \delta_x C^t = \nu \qquad \forall x \in S_\nu .$$

Show that (X_t^ν) is mixing if $\nu(S_\nu) = 1$ and

$$\lim_{t\to\infty} \delta_x P^t = \nu \qquad \forall x \in S_\nu .$$

Call $\{(\xi_t^A)\}$ nonergodic if it has more than one equilibrium. $\nu \in \mathcal{J}$ is extreme if whenever $\nu = c\nu_0 + (1-c)\nu_1$ for some $\nu_0, \nu_1 \in \mathcal{J}$ and $0 < c < 1$, then $\nu_0 = \nu = \nu_1$, i.e. if ν cannot be written as a nontrivial convex combination of invariants. According to Choquet theory, any $\nu \in \mathcal{J}$ may be written as a mixture of extremals, so for nonergodic systems one wants to find all the extreme invariant measures. Also, in this regard, μ is said to be in the domain of attraction of $\nu \in \mathcal{J}$ if μP^t (or μC^t) $\to \nu$ as $t \to \infty$. One tries to identify, as far as possible, the domain of attraction of each equilibrium.

In some cases we will be able to prove pointwise ergodic theorems. Let C be the class of continuous functions $f : S \to R$ (with the supremum norm topology). Pointwise ergodicity is an almost sure version of Cesaro convergence:

$$P\left(\frac{1}{t}\int_0^t f(\xi_s^\mu)\,ds \to \int_S f d\nu\right) = 1 \qquad \forall f \in C ,$$

for μ in the domain of attraction of $\nu \in \mathcal{J}$. In case $f(A) = A(x)$, for example, this type of result asserts that the proportion of time between 0 and t in which the site x is occupied converges almost surely to the ν-probability that x is occupied.

(1.11) Notes. The idea of constructing continuous time particle systems with the aid of percolation substructures is due to Harris (1978); more references to the origins of percolation theory can be found at the end of the next section. Examples (1.1), (1.3) and (1.5) will be studied in later sections, so we defer the relevant annotations until then. For more details on general particle systems, the structure of $\mathfrak{M}$ and $\mathcal{J}$ etc., consult the survey articles mentioned in the preface. Many of the other papers in the Bibliography deal with models which do not fit the framework of these notes; the reader is referred to those articles for a sampling of approaches to the fundamental problems of existence, uniqueness and ergodicity.

2. Percolation substructures.

In this section we construct the general percolation substructures which will be used to define particle systems. For each $x \in Z^d$, let I_x be a finite or denumerably infinite index set. Write $I = \{(i,x) : i \in I_x\}$. To each (i,x) there corresponds an "exponential alarm clock" which goes off at rate $\lambda_{i,x} \geq 0$. More precisely, let $\tau^1_{i,x}, \tau^2_{i,x}, \dots$ be an infinite sequence of increasing times such that the $\tau^{n+1}_{i,x} - \tau^n_{i,x}$, $n \geq 1$ $(\tau^0_{i,x} = 0)$, are independent exponentially distributed with rate $\lambda_{i,x}$. The graphical mechanism in $Z^d \times T$ corresponding to the (i,x) clock is determined by a set $V_{i,x} \in S_0$, and a map $W_{i,x} : Z^d \to S_0$ such that

$$(2.1) \qquad |\{y : W_{i,x}(y) \neq \{y\}\}| < \infty ,$$

as follows. First, label each point $(y, \tau^n_{i,x})$ such that $y \in V_{i,x}$ with a β (for "birth"); these will represent spontaneous sources of liquid in the percolation substructure. Second, draw a directed arrow from each $(y, \tau^n_{i,x})$ to every $(z, \tau^n_{i,x})$ such that $z \neq y$ and $z \in W_{i,x}(y)$. (No arrow emanates from $(y, \tau^n_{i,x})$ if $W_{i,x}(y) \in \{\emptyset, \{y\}\}$.) Finally, label $(y, \tau^n_{i,x})$ with a δ (for "death") if $y \notin W_{i,x}(y)$. The random graph obtained in this manner will be called the $(\lambda; V, W)$-percolation substructure, and will be denoted $\mathcal{P} = \mathcal{P}(\lambda; V, W)$. It is convenient to introduce the maps $\mathfrak{A}_{i,x} : S \to S$ given by

$$(2.2) \qquad \mathfrak{A}_{i,x}(A) = [\bigcup_{y \in A} W_{i,x}(y)] \cup V_{i,x} .$$

$\mathfrak{A}_{i,x}(A)$ may be thought of as the set of sites "wetted" by liquid either at A or spontaneously when the (i,x) clock goes off. Note that (2.1) implies $\mathfrak{A}_{i,x}(A) \Delta A \in S_0$. To ensure that the construction of $\mathcal{P}$ makes sense for our purposes, we make two assumptions on the rates: for each $y \in Z^d$,

$$(2.3) \qquad \sum_{\substack{(i,x):\\ y \in \mathfrak{A}_{i,x}(Z^d - y)}} \lambda_{i,x} < \infty ,$$

$$(2.4) \qquad \sum_{\substack{(i,x):\\ W_{i,x}(y)\neq\{y\}}} \lambda_{i,x} < \infty .$$

Condition (2.3) ensures a finite total rate at which each site is wetted either by other sites or spontaneously, while (2.4) ensures a finite total rate at which each site wets other sites or is labelled δ .

(2.5) Examples. The substructure $\mathcal{P}$ for Examples (1.1) and (1.3) has $I_x \equiv \{1,2\}$, $\lambda_{i,x} \equiv \frac{1}{2}$, $V_{i,x} \equiv \emptyset$ and $W_{i,x}$ given by $W_{1,x}(x) = \{x+1\}$, $W_{2,x}(x) = \{x-1\}$, $W_{i,x}(y) = \{y\}$ for $y \neq x$. In Example (1.5) $\mathcal{P}$ has $I_x \equiv \{0,1,2\}$, $\lambda_{0,x} \equiv 1$, $\lambda_{1,x} \equiv \lambda_{2,x} \equiv \lambda$, $V_{i,x} \equiv \emptyset$ and $W_{i,x}$ of the form: $W_{0,x}(x) = \emptyset$, $W_{1,x}(x) = \{x,x+1\}$, $W_{2,x}(x) = \{x-1,x\}$ and $W_{i,x}(y) = \{y\}$ for $y \neq x$.

A percolation substructure $\mathcal{P} = \mathcal{P}(\lambda; V, W)$ is called lineal if $V_{i,x} \equiv \emptyset$, and extralineal if $V_{i,x} \neq \emptyset$ for some $(i,x) \in I$. $\mathcal{P}$ is said to be local if there is an $L < \infty$ such that

$$(2.6) \qquad \operatorname{diam}\{y \in Z^d : y \in \mathfrak{A}_{i,x}(Z^d - y) \text{ or } W_{i,x}(y) \neq \{y\}\} \leq L \qquad \forall (i,x) \in I .$$

Condition (2.6) says that the set of sites at the tails or heads of arrows or labelled with a β or δ has diameter at most L when any clock goes off. $\mathcal{P}$ is translation invariant if $I_x \equiv I$, $\lambda_{i,x} \equiv \lambda_i$, $V_{i,x} = x + V_i$ (translate by x) and $W_{i,x}(y) \equiv W_i(y-x) + x$. Intuitively, translation invariance means that the same type of percolation mechanism applies at each site x . Note that the substructures in (2.5) are lineal, local and translation invariant.

As in Examples (1.1), (1.3) and (1.5), say there is a path up from (y,s) to (x,t) in $\mathcal{P}$ if there is a chain of alternating "upward" (= increasing in T) and "directed horizontal" (= arrowed) edges which lead from (y,s) to (x,t) without having a δ on the interior of an upward edge. By convention there is always a path up from (x,t) to (x,t) . More generally, there is a path up from D_1 to D_2 , $D_1, D_2 \in Z^d \times T$ if there is a path up from some $(y,s) \in D_1$ to some $(x,t) \in D_2$. For $t \geq 0$, set

$$\mathcal{B}_t = \{(y,s) \text{ labelled } \beta \text{ in } \mathcal{P},\ 0 < s \leq t\} , \quad \mathcal{B} = \bigcup_{t>0} \mathcal{B}_t .$$

For our construction of particle systems, a key ingredient will be the quantities

(2.7) $$N_t^A(B) = \text{the number of paths up from } (A,0) \cup \mathcal{B}_t \text{ to } (B,t) \text{ in } \mathcal{P}.$$

Given any $\mathcal{P}(\lambda; V, W)$, there is a dual substructure $\hat{\mathcal{P}}(\hat{\lambda}; \hat{V}, \hat{W})$ defined by $\hat{\lambda}_{i,x} \equiv \lambda_{i,x}$, $\hat{V}_{i,x} \equiv V_{i,x}$ and

$$z \in \hat{W}_{i,x}(y) \iff y \in W_{i,x}(z).$$

Thus the dual substructure reverses the directions of all arrows. Fix $t < \infty$, and consider $\mathcal{P}_t$ = the restriction of $\mathcal{P}$ to $Z^d \times [0,t]$. By reversing time, i.e. letting time run "down" from $\hat{0} = t$ to $\hat{t} = 0$, and reversing the direction of all arrows in $\mathcal{P}_t$, we obtain a realization of $\hat{\mathcal{P}}_t = \hat{\mathcal{P}}$ restricted to $Z^d \times [\hat{0}, \hat{t}]$ on the same probability space. This follows from the time reversibility of the sequences $(\tau^n_{i,x}; n \geq 1)$. Evidently

(2.8)
$$\begin{aligned} &\{\exists \text{ path up from } (y,s) \text{ to } (x,t) \text{ in } \mathcal{P}_t\} \\ = &\{\exists \text{ path down from } (x,\hat{0}) \text{ to } (y,\hat{t-s}) \text{ in } \hat{\mathcal{P}}_t\} \quad P-a.s., \end{aligned}$$

an observation which will be crucial for the analysis to come. The dual substructures $\hat{\mathcal{P}}_{t_0}$ corresponding to the $\mathcal{P}_{t_0}$ of figures i and ii are shown in figures iii and iv respectively.

(2.9) Problems. Let $\mathcal{P}$ be the extralineal substructure with $I_x \equiv \{1,2\}$, $\lambda_{1,x} = \kappa_x$, $\lambda_{2,x} = \lambda_x$, $V_{1,x} = \{x\}$, $V_{2,x} = \emptyset$, $W_{2,x}(x) = \emptyset$ and $W_{i,x}(y) = \{y\}$ otherwise. Put

$$\xi_t^A = \{x : N_t^A(x) > 0\}.$$

Describe the particle system $\{(\xi_t^A)\}$. Now consider the special case where $d = 1$, $\kappa_x \equiv 1$, $\lambda_x = x^2$. For this model show that if $A \in S_0$, then

(a) $\xi_t^A \in S_0$ for all $t \geq 0$ $P-a.s.$,

(b) the Markov chain (ξ_t^A) makes only instantaneous visits to each configuration $\Lambda \in S_0$.

(2.10) Problem. The basic voter model $(d = 1)$ is the spin system $\{(\xi_t^A)\}$ with flip rates

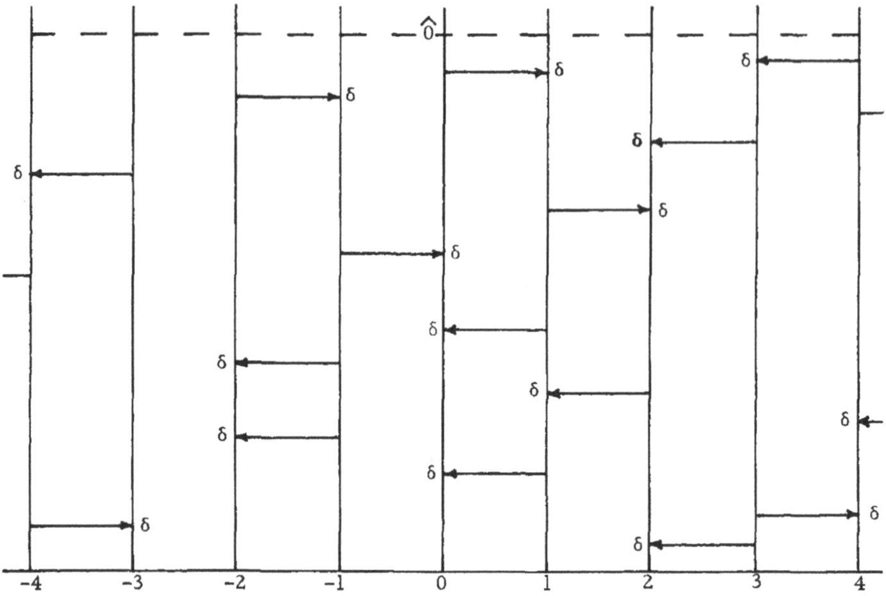

figure iii .

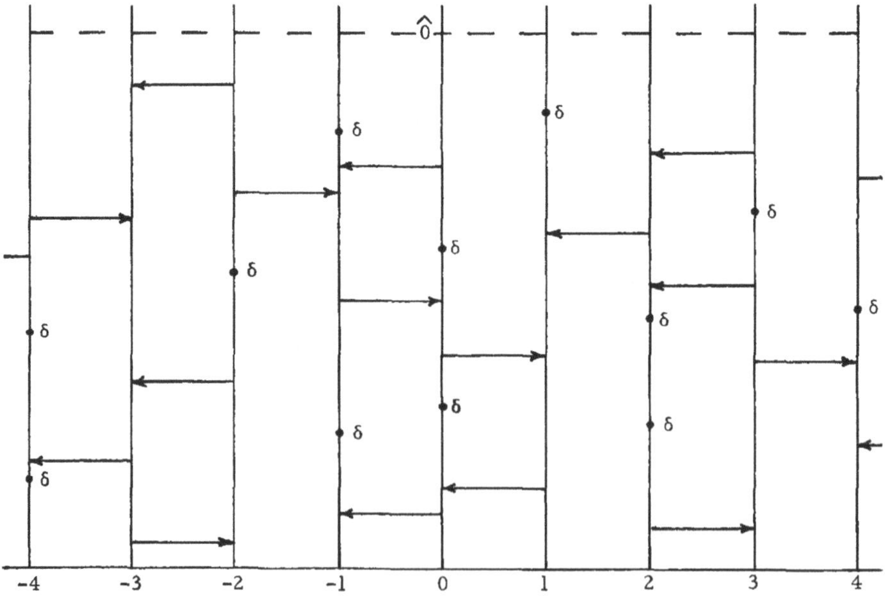

figure iv .

$$c_x(A) = \frac{1}{2}\,|A \cap \{x-1, x+1\}| \qquad x \notin A$$

$$= \frac{1}{2}\,|A^c \cap \{x-1, x+1\}| \qquad x \in A\ .$$

Show that the voter model may be defined as in (1.2), but in terms of $\hat{\wp}$, the dual substructure for the $\wp$ of Example (1.1).

(2.11) Notes. Lineal percolation substructures were introduced by Harris (1978); we refer the reader to that paper for more details of the formal construction. The idea behind this type of random graph goes back to the pioneering work on percolation by Broadbent and Hammersley (1957). For more on percolation theory and its connections with particle systems, see Clifford and Sudbury (1973), Hammersley (1959), Shante and Kirkpatrick (1971), Toom (1968) and Vasilev (1969). The instantaneous chain in Problems (2.9) is due to Blackwell (1958). References for the voter model are at the end of Section II.7.

CHAPTER II: ADDITIVE SYSTEMS

1. The general construction.

Let $\mathcal{P} = \mathcal{P}(\lambda; V, W)$ be a percolation substructure. For $t \geq 0$, $A \in S$, with $N_t^A(B)$ as in (I.2.7), define

$$\xi_t^A = \{x : N_t^A(x) > 0\} \quad . \tag{1.1}$$

Then $\{(\xi_t^A)\}$ is an S-valued Markov family, called the (canonical) additive particle system induced by $\mathcal{P}$. If $\xi_t^A = B$ and the (i,x) clock goes off, then according to (1.1), configuration B jumps to $\mathfrak{A}_{i,x}(B)$ (cf. (I.2.2)). An additive system is called lineal, extralineal, local and/or translation invariant if the underlying $\mathcal{P}$ is of the corresponding type. The term "additive" is explained by

(1.2) Proposition. If $\{(\xi_t^A)\}$ is an additive system, then

$$\xi_t^{A \cup B} = \xi_t^A \cup \xi_t^B \qquad A, B \in S, \quad t \geq 0 \quad \text{(additivity)} .$$

Proof. $N_t^{A \cup B}(x) > 0 \iff N_t^A(x) > 0$ or $N_t^B(x) > 0$. □

A particle system $\{(\xi_t^A)\}$ is called monotone if for every pair $A, B \in S$ such that $A \subset B$ there is a joint probability space on which

$$\xi_t^A \subset \xi_t^B \qquad \text{for all } t \geq 0 .$$

(1.3) Corollary. Every additive system $\{(\xi_t^A)\}$ is monotone.

Proof. By additivity, if $B \supset A$ then

$$\xi_t^B = \xi_t^A \cup \xi_t^{B-A} \supset \xi_t^A \quad \text{for all } t \geq 0 . \qquad □$$

In order to apply some of the basic facts from Chapter I, we want $\{(\xi_t^A)\}$ to be Feller. To guarantee this, one needs a very mild hypothesis on $\mathcal{P}$. Say that $\mathcal{P}$ has influence from ∞ to (x,t) if there are non-empty sets $\Lambda_1, \Lambda_2, \cdots$ and times $t_1 \geq t_2 \geq \cdots$ such that

(i) for each $n \geq 1$, there is a path up from (y, t_n) to (x, t) for all $y \in \Lambda_n$,

and

(ii) $\lim_{n\to\infty} |y_n| = \infty$ for some $y_n \in \Lambda_n$.

If, in addition, the Λ_n can be chosen so that

(iii) $\lim_{n\to\infty} |\Lambda_n| = \infty$,

then $\mathcal{P}$ is said to have strong influence from ∞ to (x, t). Influence from ∞ to (x, t) which is not strong is called weak influence.

(1.4) Proposition. If $\mathcal{P}$ is a substructure such that

$$P(\text{strong influence from } \infty \text{ to } (x, t)) = 0 \qquad \forall x \in Z^d, \qquad t \geq 0,$$

then the additive system $\{(\xi_t^A)\}$ induced by $\mathcal{P}$ is Feller.

Proof. Write $\varphi_t^A(\Lambda) = P(\xi_t^A \cap \Lambda = \emptyset)$. It suffices to show that $\varphi_t^{A_n}(\Lambda) \to \varphi_t^A(\Lambda)$ as $A_n \to A$. If $A_n \to A$, then there are $B_n \in S_0$ such that $A_n \cap B_n = A \cap B_n$ and $B_n \uparrow Z^d$ as $n \to \infty$. Now

$$\{\xi_t^{A_n} \cap \Lambda = \emptyset\} \,\Delta\, \{\xi_t^A \cap \Lambda = \emptyset\}$$
$$= \{\exists \text{ path up from exactly one of } (A_n \cap B_n^c, 0) \text{ or } (A \cap B_n^c, 0) \text{ to } (\Lambda, t)\}$$
$$\subset \{\exists \text{ path up from } (B_n^c, 0) \text{ to } (\Lambda, t)\}.$$

As $n \to \infty$, these last events converge to

$$\{\exists \text{ path up from } (B_n^c, 0) \text{ to } (\Lambda, t)\ \forall n\}$$
$$\subset \{\text{strong influence from } \infty \text{ to } (x, t) \text{ for some } x \in \Lambda\}.$$

Thus the claim follows from the hypothesis. □

We will discuss substructures with influence from ∞ in Chapter IV, but all the models in this chapter and the next will have no influence from ∞. In fact, if

(1.5) $$\sup_{y \in Z^d} \sum_{\substack{(i,x): \\ y \in \mathfrak{U}_{i,x}(Z^d - y) \\ \underline{\text{or}}\ \mathfrak{U}_{i,x}(y) = \emptyset}} \lambda_{i,x} = M < \infty ,$$

and

(1.6) $\mathcal{P}$ is local or translation invariant,

then influence from ∞ cannot occur. The specific additive systems $\{(\xi_t^A)\}$ in this chapter satisfy (1.5) and (1.6), so they are Feller by Proposition (1.4). The hypothesis of (1.4) will be assumed of general additive systems until further notice.

(1.7) Problems. Prove the above assertion that (1.5) and (1.6) imply $P(\text{influence from } \infty \text{ to } (x,t)) = 0 \ \forall x,t$. Give an example of a substructure $\mathcal{P}$ with strong influence from ∞ for which the canonical additive system $\{(\xi_t^A)\}$ defined by (1.1) is not Feller.

We now proceed to derive the duality equation for additive systems. Adjoin an isolated point Δ to S_0, and write $\hat{S} = S_0 \cup \Delta$. By convention, $\Delta \cap A \neq \emptyset$ for all $A \in S$ ($\emptyset$ included!) Let $\hat{\mathcal{P}}$ be the dual substructure for $\mathcal{P}$, and define times $\hat{\tau}_\Delta^B$, $B \in S_0$, on $\hat{\mathcal{P}}$ by

$$\hat{\tau}_\Delta^B = \inf\{t \geq 0 : \exists \text{ path up from } (B,0) \text{ to } \mathcal{B}_t \text{ in } \hat{\mathcal{P}}\}$$
$$(= \infty \text{ if no such } t \text{ exists}).$$

Next, introduce the family $\{(\hat{\xi}_t^B);\ B \in S_0\}$ of $\hat{S}$-valued Markov chains, called the dual processes for $\{(\xi_t^A)\}$, and given by

$$\hat{\xi}_t^B = \{x : \exists \text{ path up from } (B,0) \text{ to } (x,t) \text{ in } \hat{\mathcal{P}}\} \quad t < \hat{\tau}_\Delta^B$$
$$= \Delta \quad t \geq \hat{\tau}_\Delta^B .$$

Finally, introduce

$$\hat{\tau}_\emptyset^B = \inf\{t \geq 0 : \hat{\xi}_t^B = \emptyset\} \quad (= \infty \text{ if no such } t \text{ exists}).$$

Note that $\emptyset$ and Δ are both traps for $(\hat{\xi}_t^B)$, so at most one of $\hat{\tau}_\Delta^B$ and $\hat{\tau}_\emptyset^B$ is finite. Our first theorem will be the main tool in the study of additive systems.

To state it, we introduce the notation

$$\varphi_t^{\mu} = \varphi^{\mu P^t}, \quad \varphi_t^{A} = \varphi^{\delta_A P^t}, \quad \hat{\varphi}_t^{B}(A) = \hat{P}(\hat{\xi}_t^{B} \cap A = \emptyset),$$

$\mu \in \mathfrak{M}$, $A \in S$, $B \in S_0$.

(1.8) <u>Theorem</u> (<u>Additive duality equation</u>.) Let $\{(\xi_t^A)\}$ be the additive system induced by $\mathcal{P}$, $\{(\hat{\xi}_t^B);\ B \in S_0\}$ the corresponding dual system. For each $t \geq 0$, $A \in S$, $B \in S_0$,

(1.9) $$\varphi_t^{A}(B) = \hat{\varphi}_t^{B}(A) .$$

More generally, if $\hat{E}$ is the expectation operator for $\hat{P}$,

(1.10) $$\varphi_t^{\mu}(B) = \hat{E}[\varphi^{\mu}(\hat{\xi}_t^B)] \qquad t \geq 0, \quad \mu \in \mathfrak{M}, \quad B \in S_0 .$$

For μ_θ = Bernoulli product measure with density θ,

(1.11) $$\varphi_t^{\mu_\theta}(B) = \hat{E}[(1-\theta)^{|\hat{\xi}_t^B|}] .$$

<u>Proof</u>. Let $\mathcal{P}_t$ and $\hat{\mathcal{P}}_t$ be the forward and reverse percolation substructures on $Z^d \times [0,t] = Z^d \times [\hat{0},\hat{t}]$ which were discussed toward the end of Chapter I. According to (I.2.8), copies of $(\xi_s^A)_{0 \leq s \leq t}$ and $(\hat{\xi}_s^B)_{0 \leq s \leq t}$ are both defined by these joint substructures in such a way that

$$\{\xi_t^A \cap B = \emptyset\} = \{\hat{\xi}_t^B \cap A = \emptyset\} \qquad P\text{ - a.s.}$$

In fact, both events are P - a.s. equivalent to

$$\{\nexists \text{ a path between } (A,0) \cup \mathcal{B}_t \text{ and } (B,t)\} .$$

The first duality equation (1.9) follows. To get (1.10), integrate (1.9) with respect to μ :

$$\begin{aligned}\varphi_t^{\mu}(B) &= \int \varphi_t^{A}(B)\mu(dA) \\ &= \int \hat{\varphi}_t^{B}(A)\mu(dA) \\ &= \int \sum_{\Lambda : A\cap\Lambda=\emptyset} \hat{P}(\hat{\xi}_t^{B} = \Lambda)\,\mu(dA) \\ &= \sum_{\Lambda\in S_0}' \hat{P}(\hat{\xi}_t^{B} = \Lambda)\varphi^{\mu}(\Lambda) = \hat{E}[\varphi^{\mu}(\hat{\xi}_t^{B})] .\end{aligned}$$

Since $\varphi^{\mu_\theta}(\Lambda) = (1-\theta)^{|\Lambda|}$, (1.11) follows from (1.10). □

As easy consequences of the duality equation, we obtain two results on the ergodic theory of general additive systems.

(1.12) Corollary. Let $\{(\xi_t^A)\}$ be additive. There are extreme invariant measures ν_0 and ν_1 such that

$$\delta_{\emptyset} P^t \to \nu_0 \quad \text{and} \quad \delta_{Z^d} P^t \to \nu_1 \quad \text{as} \quad t\to\infty .$$

The zero functions for ν_0 and ν_1 are

(1.13) $$\varphi^{\nu_0}(\Lambda) = \hat{P}(\hat{\tau}_{\Delta}^{\Lambda} = \infty), \quad \varphi^{\nu_1}(\Lambda) = \hat{P}(\hat{\tau}_{\emptyset}^{\Lambda} < \infty) .$$

Proof. According to the remarks on convergence in Chapter I, it suffices to note that for any $\Lambda \in S_0$,

$$\varphi_t^{\emptyset}(\Lambda) = \hat{P}(\hat{\tau}_{\Delta}^{\Lambda} > t) \downarrow \hat{P}(\hat{\tau}_{\Delta}^{\Lambda} = \infty)$$

and

$$\varphi_t^{Z^d}(\Lambda) = \hat{P}(\hat{\tau}_{\emptyset}^{\Lambda} \le t) \uparrow \hat{P}(\hat{\tau}_{\emptyset}^{\Lambda} < \infty)$$

as $t\to\infty$. The fact that ν_0 and ν_1 are extreme follows from an easy monotonicity argument (cf. Problem (1.14)). □

(1.14) Problem. Use Corollary (1.3) to show that if $\mu \in \mathcal{J}$, then $\varphi^{\nu_0} \ge \varphi^{\mu} \ge \varphi^{\nu_1}$. Conclude that ν_0 and ν_1 are extreme equilibria.

(1.15) Corollary. The additive system $\{(\xi_t^A)\}$ is ergodic if and only if

(1.16) $$\hat{P}(\hat{\tau}_{\emptyset}^{\Lambda} \wedge \hat{\tau}_{\Delta}^{\Lambda} < \infty) = 1 \qquad \forall \Lambda \in S_0 .$$

($a \wedge b$ denotes the minimum of a and b.)

Proof. From the previous corollary, we see that (1.16) is equivalent to $\nu_0 = \nu_1$. By Corollary (1.3), on the other hand, $\xi_t^{\emptyset} \subset \xi_t^A \subset \xi_t^{Z^d}$ so that $\varphi_t^{\emptyset} \geq \varphi_t^A \geq \varphi_t^{Z^d}$ for all $t \geq 0$, $A \in S$. If $\nu_0 = \nu_1 = \nu$, say, we conclude that $\varphi_t^A \to \varphi_t^{\nu}$, establishing ergodicity. □

A few words about additive processes starting from arbitrary $\mu \in \mathfrak{M}$ are in order at this point. Note that while (1.1) defines (ξ_t^A) simultaneously for all A, it does not represent (ξ_t^{μ}) for nondeterministic initial distributions. When μ is prescribed, a straightforward way to get such a representation is to enlarge the underlying probability space to support a μ-distributed random subset γ and set $\xi_t^{\mu} = \xi_t^A$ on $\{\gamma = A\}$. On occasions where we consider such a process (ξ_t^{μ}), it will be assumed without further comment that this construction has been carried out.

(1.17) Notes. Lineal additive processes are studied by Harris (1978); see also Bertein and Galves (1978). Graphical duality has appeared in one form or another in Broadbent and Hammersley (1957), Clifford and Sudbury (1973), Harris (1978), Toom (1968) and Vasilev (1969). For another more analytical approach to duality, the reader is referred to Harris (1976), Holley and Liggett (1975), Holley and Stroock (1976d), Holley, Stroock and Williams (1977) and Vasershtein and Leontovich (1970). Monotone (= attractive) systems are discussed by Holley (1972b).

2. Ergodic theorems for extralineal additive systems.

In this section we derive general ergodic theorems for extralineal additive systems. A particle system $\{(\xi_t^A)\}$ is called exponentially ergodic if there is a measure $\nu \in \mathfrak{M}$ and a constant $\alpha > 0$ such that, for every $\mu \in \mathfrak{M}$, $\Lambda_0 \subset \Lambda \in S_0$,

(2.1) $$|\mu P^t(\{A : A \cap \Lambda = \Lambda_0\}) - \nu(\{A : A \cap \Lambda = \Lambda_0\})| \leq c_{\Lambda} e^{-\alpha t} ,$$

where c_{Λ} is a positive constant depending only on Λ.

(2.2) <u>Theorem</u>. Let $\{(\xi_t^A)\}$ be an extralineal additive system with substructure $\mathcal{P}(\lambda; V, W)$. If

$$\inf_{y \in Z^d} \sum_{\substack{(i,x): \\ y \in V_{i,x}}} \lambda_{i,x} = \kappa > 0 , \tag{2.3}$$

then the system is exponentially ergodic. In fact, (2.1) holds with $\alpha = \kappa$.

<u>Proof</u>. Condition (2.3) states that a β appears in the substructures $\mathcal{P}$ and $\hat{\mathcal{P}}$ at each site y with rate at least $\kappa > 0$. Thus, from any non-empty finite Λ the dual process goes to Δ with rate at least κ. It follows that

$$\hat{P}(\hat{\tau}_\emptyset^B \wedge \hat{\tau}_\Delta^B > t) \leq e^{-\kappa t} \qquad B \in S_0, \quad t \in T. \tag{2.4}$$

By duality, for any $A \in S$, $B \in S_0$,

$$\varphi_t^A(B) = \hat{P}(\hat{\tau}_\emptyset^B \leq t) + \hat{P}(\hat{\xi}_t^B \cap A = \emptyset, \ t < \hat{\tau}_\emptyset^B \wedge \hat{\tau}_\Delta^B).$$

Rearranging,

$$\varphi_t^A(B) - \hat{P}(\hat{\tau}_\emptyset^B < \infty)$$
$$= \hat{P}(\hat{\xi}_t^B \cap A = \emptyset, \ t < \hat{\tau}_\emptyset^B \wedge \hat{\tau}_\Delta^B) - \hat{P}(t < \hat{\tau}_\emptyset^B < \infty) .$$

Now apply (2.4) to get

$$|\varphi_t^A(B) - \varphi^{\nu_1}(B)| \leq 2 e^{-\kappa t} ,$$

and use inclusion-exclusion to finish the proof of (2.1). □

(2.5) <u>Corollary</u>. Any translation invariant extralineal additive system is exponentially ergodic.

<u>Proof</u>. (2.3) is automatic in the translation invariant case. □

Our next result asserts that in many cases the unique invariant ν in Theorem (2.2) has <u>exponentially decaying correlations</u>.

(2.6) Theorem. Given a local extralineal substructure $\mathcal{P}$ which satisfies (1.5) and (2.3), let $\{(\xi_t^A)\}$ be the additive system induced by $\mathcal{P}$. If ν is the unique equilibrium for $\{(\xi_t^A)\}$ which is guaranteed by Theorem (2.2), then

$$(2.7) \qquad |\varphi^\nu(B \cup C) - \varphi^\nu(B)\varphi^\nu(C)| \le c e^{-\alpha d(B,C)} \qquad B, C \in S_0,$$

where c and α are constants depending only on $(\lambda; V, W)$, and $d(B,C) = \min\{|x-y| : x \in B,\ y \in C\}$ is the distance between B and C.

Proof. Fix $B, C \in S_0$. By (1.13), the left side of (2.7) equals

$$(2.8) \qquad |\hat{P}(\hat{\tau}_\emptyset^{B \cup C} < \infty) - \hat{P}(\hat{\tau}_\emptyset^B < \infty)\hat{P}(\hat{\tau}_\emptyset^C < \infty)| .$$

Let $\hat{\mathcal{P}}_1$ and $\hat{\mathcal{P}}_2$ be two independent copies of $\hat{\mathcal{P}}$. Use $\hat{\mathcal{P}}_1$ to define $(\hat{\xi}_t^B)$ and $\mathcal{P}_2$ to define $(\hat{\xi}_t^C)$; note that this is not the standard graphical representation, since the two processes are independent. With L as in (I.2.6), define

$$\hat{\tau}_L = \min\{t : d(\hat{\xi}_t^B, \hat{\xi}_t^C) \le L\} \qquad (= \infty \text{ if no such } t \text{ exists}).$$

Now manufacture a copy of $(\hat{\xi}_t^{B \cup C})$ by letting the flow starting from B use $\hat{\mathcal{P}}_1$, while the flow starting from C uses $\mathcal{P}_2$ until $\hat{\tau}_L$ and $\mathcal{P}_1$ thereafter. Under this representation, $\hat{P}(\hat{\tau}_\emptyset^B < \infty,\ \hat{\tau}_\emptyset^C < \infty) = \hat{P}(\hat{\tau}_\emptyset^B < \infty)\hat{P}(\hat{\tau}_\emptyset^C < \infty)$, and $\hat{\xi}_t^{B \cup C} = \hat{\xi}_t^B \cup \hat{\xi}_t^C$ on $\{\hat{\tau}_L \ge t\}$. Thus

$$\{\hat{\tau}_\emptyset^{B \cup C} < \infty\} \,\Delta\, \{\hat{\tau}_\emptyset^B < \infty,\ \hat{\tau}_\emptyset^C < \infty\} \subset \{\hat{\tau}_L < \hat{\tau}_\Delta^{B \cup C}\},$$

so that (2.8) is majorized by $\hat{P}(\hat{\tau}_L < \hat{\tau}_\Delta^{B \cup C})$. Observe that there must be at least $\lfloor \frac{d(B,C)}{L} \rfloor$ jumps by $(\hat{\xi}_t^{B \cup C})$ by time $\hat{\tau}_L$ ($\lfloor r \rfloor$ denotes the greatest integer less than r.) Conditions (1.5) and (2.3) ensure that at each jump time $(\hat{\xi}_t^{B \cup C})$ goes to Δ with probability at least $p = \frac{\kappa}{M} > 0$. Assuming $p < 1$ (the theorem is trivial if $p = 1$), we conclude that

$$\hat{P}(\hat{\tau}_L < \hat{\tau}_\Delta^{B \cup C}) \le \hat{P}(\lfloor \tfrac{d(B,C)}{L} \rfloor \text{ jumps occur before } \tau_\Delta)$$
$$\le (1-p)^{\lfloor \frac{d(B,C)}{L} \rfloor} < (1-p)^{-1} e^{-\frac{p}{L} d(B,C)},$$

as desired. □

When $\mathcal{P}$ satisfies (2.3), the induced additive system is also pointwise ergodic, in the sense described toward the end of Section I.1.

(2.9) Theorem. Let $\{(\xi_t^A)\}$ satisfy the hypotheses of Theorem (2.2), and let ν be its unique equilibrium. Then for each $f \in \mathcal{C}$, $\mu \in \mathcal{M}$,

(2.10) $$P\left(\lim_{t \to \infty} \frac{1}{t} \int_0^t f(\xi_s^\mu)\, ds = \int_S f\, d\nu \right) = 1 .$$

In particular, (2.10) holds for any translation invariant extralineal additive system.

Proof. Define ξ_t^ν by the procedure outlined at the end of Section II.1. Problem (I.1.10) implies that the stationary process (ξ_t^ν) is Birkhoff ergodic, and in fact mixing. By Birkhoff's Theorem, it follows that (2.10) holds in case $\mu = \nu$, for any $f \in L^1(\nu)$. Now for each $x \in Z^d$, $A, B \in S$, the duality construction and (2.4) show that

$$P(\xi_t^A(x) \neq \xi_t^B(x)) \leq P(N_t^{\emptyset}(x) = 0,\; N_t^{A \cup B}(x) > 0)$$
$$\leq \hat{P}(\hat{\tau}_{\emptyset}^x \wedge \hat{\tau}_{\Delta}^x > t) \leq e^{-\kappa t} .$$

By a routine Borel-Cantelli argument,

$$P(\xi_t^A(x) = \xi_t^B(x) \;\; \forall \text{ sufficiently large } t) = 1 .$$

Hence

$$P(\xi_t^A(x) = \xi_t^\nu(x) \;\; \forall \text{ sufficiently large } t) = 1 \qquad x \in Z^d,\; A \in S .$$

Let $\mathfrak{F}$ be the class of functions $f : S \to R$ which depend on only finitely many sites (the so-called tame or cylinder functions.) For $f \in \mathfrak{F}$ we conclude that

$$\tau_f = \min\{t : f(\xi_s^A) = f(\xi_s^\nu) \;\forall\; s \geq t\} < \infty \qquad P\text{ - a.s.}$$

To finish the proof of (2.9) for $f \in \mathfrak{F}$, $\mu = \delta_A$, note that

$$\frac{1}{t} \int_0^t f(\xi_s^A)\, ds = \frac{1}{t} \int_0^{\tau_f \wedge t} f(\xi_s^A)\, ds + \frac{1}{t} \int_{\tau_f \wedge t}^t f(\xi_s^A)\, ds$$
$$\to \lim_{t \to \infty} \frac{1}{t} \int_{\tau_f}^t f(\xi_s^\nu)\, ds = \int f\, d\nu$$

as $t \to \infty$, the last equality by virtue of (2.9) for $\mu = \nu$. The extensions from $\mathfrak{F}$ to C and from δ_A to general μ are accomplished by means of easy approximation arguments. □

(2.11) <u>Problem</u>. Give an example showing that under the hypotheses of the last theorem, pointwise ergodicity need not hold for general $f \in L^1(\nu)$ when the process starts from arbitrary $A \in S$.

Our next task will be to identify the class of additive <u>spin systems</u>, in particular those for which Theorems (2.2), (2.6) and (2.9) apply. If a process can only change configuration at one site at a time, then necessarily $|V_{i,x}| \leq 1$ for all (i,x). Without loss of generality we can assume

$$\lambda_{0,x} = \kappa_x, \quad V_{0,x} = \{x\}, \quad V_{i,x} = \emptyset \quad \text{otherwise.}$$

Also, the W's can be chosen to be of the form

$$W_{0,x}(y) = \{y\} \quad \forall y,$$

and for $i \neq 0$,

$$\begin{aligned} W_{i,x}(y) &= \{x,y\} & & y \in C_{i,x} \\ &= \{y\} & & y \notin C_{i,x}, \quad y \neq x \\ &= \emptyset & & y = x \notin C_{i,x}, \end{aligned}$$

for some $C_{i,x} \in S_0$. Thus $\mathcal{P}$ is determined by (i) rates $\kappa_x \geq 0$ with which a β appears at x, and (ii) rates $\lambda_{i,x}$ and finite sets $C_{i,x}$, $i \neq 0$, such that when the (i,x) clock goes off an arrow is directed from every site of $C_{i,x} - \{x\}$ to x and x is labelled with a δ if $x \notin C_{i,x}$. It is convenient to redefine I_x, by removing $i = 0$. Then if our process is in state A, with $x \notin A$, a flip at x occurs with rate

$$\kappa_x + \sum_{\substack{i \in I_x: \\ A \cap C_{i,x} \neq \emptyset}} \lambda_{i,x},$$

while if $x \in A$ a flip occurs at x with rate

$$\sum_{\substack{i \in I_x : \\ A \cap C_{i,x} = \emptyset}} \lambda_{i,x} \quad .$$

Conditions (I.2.3) and (I.2.4) are covered by the requirement that

$$\lambda_x = \sum \lambda_{i,x} < \infty \qquad \forall x \in Z^d .$$

The flip rates may be consolidated in the form

(2.12) $$c_x(A) = \kappa_x(1-A(x)) + \lambda_x A(x) + (1-2A(x)) \sum_{\substack{i \in I_x : \\ A \cap C_{i,x} \neq \emptyset}} \lambda_{i,x} \quad .$$

Condition (1.5) is equivalent to

(2.13) $$\sup_{x,A} c_x(A) < \infty \quad .$$

A particle system $\{(\xi_t^A)\}$ is called a proximity system if its flip rates have the form (2.12) for some $\kappa_x \geq 0$, $\lambda_{i,x} \geq 0$ and $C_{i,x} \in S_0$. It is lineal if $\kappa_x \equiv 0$, extralineal otherwise. The local property is

$$\sup_x \operatorname{diam}[\{x\} \cup (\bigcup_{i \in I_x} C_{i,x})] < \infty \quad .$$

In the translation invariant case $\kappa_x \equiv \kappa$, $I_x \equiv I$, $C_{i,x} \equiv x + C_i$ and $\lambda_{i,x} \equiv \lambda_i$. The hypothesis (2.3) holds if $\inf_x \kappa_x > 0$.

(2.14) Problems. Check the various assertions made about proximity systems. In particular, verify that every additive spin system is a proximity system. Show by example that distinct substructures $\mathcal{P}_1$ and $\mathcal{P}_2$ can give rise to the same additive system, i.e. that distinct $(\lambda_1; V_1, W_1)$ and $(\lambda_2; V_2, W_2)$ can induce the same jump rates. Prove that any additive system has a representation in terms of a substructure $\mathcal{P}(\lambda; V, W)$ such that either $V_{i,x} = \emptyset$ or $W_{i,x}(y) = \{y\}\ \forall y$.

If $\{(\xi_t^A)\}$ is a proximity system, then its dual processes $(\hat{\xi}_t^B)$ on $\hat{S}$ are coalescing branching processes. At rate $\lambda_{i,x}$ a particle in the dual tries to replace itself with particles situated on $C_{i,x} \in S_0$. Whenever two particles attempt to occupy the same site they coalesce into one. At rate κ_x a particle at x sends the whole process to Δ. In keeping with Corollary (1.15), ergodicity of the proximity system is equivalent to eventual absorption of the corresponding coalescing branching system at either $\emptyset$ or Δ with probability one.

(2.15) <u>Problem</u>. Let $\{(\xi_t^A)\}$ be an extralineal proximity process such that $\kappa_x + \lambda_x > 0$ for all x, and

$$\inf_x \frac{\kappa_x}{\kappa_x + \lambda_x} > 0 \ .$$

Prove that the system is (strongly) ergodic. Show by example that the convergence need not be exponential.

(2.16) <u>Problem</u>. Let $\{(\xi_t^A)\}$ be a (one dimensional) basic voter model with spontaneous birth at the origin, i.e. the extralineal proximity system with flip rates

$$c_0(A) = \kappa(1-A(0)) + A(0) + (\tfrac{1}{2} - A(0))|A \cap \{-1,1\}| \ ,$$

$$c_x(A) = A(x) + (\tfrac{1}{2} - A(x))|A \cap \{x-1,x\}| \quad \cdot \ x \neq 0 \ ,$$

for some $\kappa > 0$. Prove that the system is (strongly) ergodic.

The final result of this section is a correlation inequality for proximity systems.

(2.17) <u>Theorem</u>. If $\{(\xi_t^A)\}$ is a proximity system, then

$$\varphi_t^A(B \cup C) \geq \varphi_t^A(B)\,\varphi_t^A(C) \qquad A \in S, \ B, C \in S_0, \ t \in T \ .$$

<u>Proof</u>. By duality, it suffices to check the equivalent inequalities

$$\hat{\varphi}_t^{B \cup C}(A) \geq \hat{\varphi}_t^B(A)\,\hat{\varphi}_t^C(A) \ . \tag{2.18}$$

To do this, we use a strategy similar to the one which proved Theorem (2.6). Namely, we fix B and C, and construct <u>independent</u> copies of $(\hat{\xi}_t^B)$ and $(\hat{\xi}_t^C)$ by using independent substructures $\hat{\mathcal{P}}_1$ and $\hat{\mathcal{P}}_2$ to define them. But now we introduce a different representation of $(\hat{\xi}_t^{B \cup C})$, by making use of the coalescing branching process interpretation. Namely, whenever a particle from $(\hat{\xi}_t^B)$ collides with one from $(\hat{\xi}_t^C)$, the former survives and the latter dies. Since this collision mechanism is indistinguishable from coalescence, we do in fact obtain a copy of

$(\hat{\xi}_t^{B \cup C})$ with the key property

(2.19) $$\hat{\xi}_t^{B \cup C} \subset \hat{\xi}_t^B \cup \hat{\xi}_t^C \qquad \forall t \in T .$$

In terms of our construction, (2.18) is equivalent to

$$P(\hat{\xi}_t^{B \cup C} \cap A = \emptyset) \geq P((\hat{\xi}_t^B \cup \hat{\xi}_t^C) \cap A = \emptyset) ,$$

an immediate consequence of (2.19). □

(2.20) Problems. Show by example that the correlation inequalities of the last theorem do not hold for all additive systems. For which additive $\{(\xi_t^A)\}$ other than proximity systems are the inequalities valid?

(2.21) Notes. A result closely related to Theorem (2.2) may be found in Schwartz (1977). For versions of (2.2) in the spin system setting, see Holley and Stroock (1976d) and (in discrete time) Vasershtein and Leontovich (1970). The discrete time analogue of Theorem (2.6) is proved by Bramson and Griffeath (1978a); similar but more sophisticated inequalities for the stochastic Ising model (cf. (III.3)) have been obtained by Holley and Stroock (1976b). R. Arratia (private communication) has shown that ν satisfies a strong form of exponential mixing when the hypotheses of (2.6) are satisfied. Pointwise ergodic theorems for particle systems were first obtained by Harris (1978); we note that Theorem (2.8) can also be proved by generalizing the criterion he gives for lineal additive systems. Lineal proximity systems and coalescing branching processes were introduced by Holley and Liggett (1975). Problems (2.15) and (2.16) are adapted from Holley and Stroock (1976a) and Schwartz (1977) respectively. Harris (1977) has proved a much more general version of Theorem (2.17) by an entirely different method.

3. Lineal additive systems.

If the percolation substructure $\mathcal{P}(\lambda; V, W)$ is lineal, i.e. if $V_{i,x} \equiv \emptyset$ so that no β's appear, then we abbreviate $\mathcal{P} = \mathcal{P}(\lambda, W)$. Additive systems induced by lineal substructures have the important property that spontaneous creation is impossible. In other words, $\emptyset$ is a trap so that $\delta_\emptyset$ is invariant. In biological contexts such systems might be termed "biogenetic" (as opposed to "abiogenetic"). Ergodicity is therefore equivalent to weak convergence to $\delta_\emptyset$ from any initial state, and the ergodic theory of lineal systems turns out to be much more delicate than that of extralineal ones. The remaining sections of this chapter will be devoted to the study of specific lineal additive systems (e.g. contact processes, voter models, coalescing random walks) in some detail. But first, we note a few simplifications which take place in the duality theory for the lineal case, and prove an ergodic theorem for lineal proximity processes.

(3.1) <u>Theorem</u>. Let $\{(\xi_t^A)\}$ be the lineal additive system induced by a substructure $\mathcal{P}(\lambda, W)$, $\{(\hat{\xi}_t^B); B \in S\}$ the lineal additive system induced by the dual substructure $\hat{\mathcal{P}}(\lambda, \hat{W})$. Let φ_t^A and $\hat{\varphi}_t^B$ be the zero functions of the respective systems. For each $t \in T$, $A, B \in S$,

$$\varphi_t^A(B) = \hat{\varphi}_t^B(A) . \tag{3.1}$$

There is an extreme invariant measure $\nu_1 \in \mathfrak{M}$ such that $\delta_{Z^d} \to \nu_1$ as $t \to \infty$. Moreover,

$$\{(\xi_t^A)\} \text{ ergodic} \iff \nu_1 = \delta_\emptyset$$

$$\iff \hat{P}(\hat{\tau}_\emptyset^\Lambda < \infty) = 1 \qquad \forall \Lambda \in S_0 .$$

<u>Proof</u>. In the lineal case $\mathcal{B} = \emptyset$, $\hat{\tau}_\Delta^B = \infty$ P - a.s. for all $B \in S_0$, and so $(\hat{\xi}_t^B)$ is simply the process induced by $\hat{\mathcal{P}}$. In this case $(\hat{\xi}_t^B)$ can be defined for <u>all</u> $B \in S$, and (3.1) holds because on the joint substructure the events $\{\xi_t^A \cap B = \emptyset\}$ and $\{\hat{\xi}_t^B \cap A = \emptyset\}$ are both P - a.s. equivalent to

$$\{\nexists \text{ path between } (A,0) \text{ and } (B,t)\} .$$

The remaining assertions are simply the specialization of Corollaries (1.12) and (1.15) to the lineal case. □

If $\{(\xi_t^A)\}$ is a lineal proximity system, then its dual $\{(\hat{\xi}_t^A)\}$ is the lineal system of coalescing branching processes determined by the λ_{ix}'s and $\hat{W}_{i,x}$'s of the form

$$W_{i,x}(y) = C_{i,x} \qquad y = x$$
$$= \{y\} \qquad y \neq x .$$

An ergodic theorem for lineal proximity systems is easily obtained by exploiting the branching interpretation of their duals.

(3.2) Theorem. Let $\{(\xi_t^A)\}$ be a lineal proximity system, with flip rates of the form (2.12) (with $\kappa_x \equiv 0$). Set

$$\iota = \inf_{x \in Z^d} \lambda_x , \quad m = \sup_{\substack{x \in Z^d: \\ \lambda_x > 0}} \sum_{i \in I_x} \frac{\lambda_{i,x}}{\lambda_x} |C_{i,x}| .$$

If $\iota > 0$ and $m < 1$, then the system is exponentially ergodic. In the translation invariant case the system is also ergodic if $m = 1$ and $C_{i,x} = C_i = \emptyset$ for some i; in fact,

(3.3) $$1 - \varphi_t^{\mu}(\Lambda) \leq c|\Lambda| t^{-1} \qquad \mu \in \mathfrak{M} , \quad t \in T , \quad \Lambda \in S_0 ,$$

where c is a positive constant depending only on (λ, W).

Sketch of proof: The dual process $(\hat{\xi}_t^{\Lambda})$ can be imbedded in a spatial branching process $(\tilde{\xi}_t^{\Lambda})$, where the latter ignores the coalescence mechanism. The conditions $\iota > 0$, $m < 1$ ensure that $(\hat{\xi}_t^{\Lambda})$ dies out more quickly than a subcritical Galton-Watson process, and since $\hat{\xi}_t^{\Lambda} \subset \tilde{\xi}_t^{\Lambda}$, the exponential ergodicity of $\{(\xi_t^{\Lambda})\}$ follows. In the translation invariant case $(|\tilde{\xi}_t^{\Lambda}|)$ is a Galton-Watson process, so (3.3) is a consequence of the same comparison and a well-known rate of extinction result for nontrivial critical branching processes. We leave the details of the proof, and some complementary results, as an exercise.

(3.4) Problems. Fill in the details of the last proof. Show that if $\lambda_x > 0$ for all $x \in Z^d$ and

(a) $m < 1$,

or

(b) $$\inf_{x \in Z^d} \sum_{\substack{i \in I_x: \\ C_{i,x} = \emptyset}} \lambda_{i,x} > 0 \quad \text{and} \quad m = 1 ,$$

then $\{(\xi_t^A)\}$ is still ergodic.

(3.5) Problems. A lineal substructure $\mathcal{P}(\lambda, W)$ is called self-dual if $\hat{W} = W$ so that $\hat{\mathcal{P}}$ coincides with $\mathcal{P}$. A lineal additive system is self-dual if its substructure is self-dual. Show that the basic contact systems of (I.1.5) are self-dual. Thus these systems may be thought of as either proximity systems or coalescing branching systems. Apply Theorem (3.2) to prove ergodicity for any $\lambda \le \frac{1}{2}$. Find other examples of self-dual additive systems.

(3.6) Notes. Lineal additive systems are discussed, in varying degrees of generality, by Harris (1976, 1978), Holley and Liggett (1975), Holley and Stroock (1976d), and (in discrete time) Vasershtein and Leontovich (1970). The prototype for Theorem (3.2) is in Holley and Liggett (1975); see Liggett (1977) for a more detailed presentation of (3.2) and (3.3).

4. Contact systems: basic properties.

In this section we study two families of translation invariant lineal proximity systems in considerable detail: the basic (one-dimensional) contact systems, and the one-sided (one-dimensional) contact systems. The former were described in Example (I.1.5); recall that they were formulated as models for the spread of an infection. When we want to exhibit the dependence on the infection parameter λ, we will write $\xi_t^A = \xi_{\lambda,t}^A$. The one-sided model is defined by taking $I_x \equiv \{0,1\}$, $\lambda_{0,x} \equiv 1$, $\lambda_{1,x} \equiv \lambda$, $C_{0,x} \equiv \emptyset$ and $C_{1,x} = \{x-1, x\}$ in (2.12) ($\kappa_x \equiv 0$). Schematically, $\mathcal{P}$ has:

δ at each x with rate 1

$\rightarrow$ from $x-1$ to x with rate λ.

Thus a site can only be infected by its left neighbor, so the infection spreads to the right. Denote these systems as $\{(\xi_t^{+,A})\} = \{(\xi_{\lambda,t}^{+,A})\}$, $\lambda \ge 0$.

The main feature of contact systems is that they exhibit a critical phenomenon: there is a λ_*, $0 < \lambda_* < \infty$, such that ergodicity (= recovery) occurs when the infection rate λ is less than λ_*, while infection persists forever if $\lambda > \lambda_*$. To distinguish between the two models, which have different critical values, the two-sided value will be denoted by λ_*, the one-sided value by λ_*^+. Our first tasks will be to prove that the critical phenomenon takes place, and to obtain bounds on λ_* and λ_*^+.

(4.1) Proposition. Let $\{(\xi_{\lambda,t}^A)\}$ be the basic contact system with parameter λ, and write

$$p_{\lambda,t} = P(0 \in \xi_{\lambda,t}^Z) \quad ,$$

(4.2) $$p_\lambda = \lim_{t\to\infty} p_{\lambda,t} = \nu_{\lambda,1} \quad (\{0 \text{ is infected}\})$$

$(\nu_{\lambda,1} = \lim_{t\to\infty} \delta_Z P_\lambda^t)$. Then

(4.3) $$p_{\lambda_1,t} \le p_{\lambda_2,t} \quad \text{if} \quad \lambda_1 \le \lambda_2, \quad t \in T,$$

and so p_λ is an increasing function of λ. If

$$\lambda_* = \sup\{\lambda : p_\lambda = 0\} ,$$

the systems with $\lambda < \lambda_*$ are ergodic, while those with $\lambda > \lambda_*$ are nonergodic. Letting $p_{\lambda,t}^+$, p_λ^+ and λ_*^+ be defined similarly in terms of the one-sided systems $\{(\xi_{\lambda,t}^{+,A})\}$, the analogous assertions hold.

Proof. Property (4.2) follows from Theorem (3.1). To show (4.3), fix $0 \le \lambda_1 < \lambda_2 < \infty$. Let $\mathcal{P}$ be the substructure of Example (I.1.5) with $\lambda = \lambda_1$, and define $\xi_{\lambda_1,t}^A$ in terms of $\mathcal{P}$ by (1.1). Now augment $\mathcal{P}$ by adding additional arrows of the same types which arrive at each site x with rates $\lambda_2 - \lambda_1 > 0$; call the augmented substructure $\overline{\mathcal{P}}$. The system $\{(\xi_{\lambda_2,t}^A)\}$ can be represented in terms of $\overline{\mathcal{P}}$ via (1.1). Observe that

$$\xi_{\lambda_1,t}^A \subset \xi_{\lambda_2,t}^A \qquad A \in S, \quad t \in T,$$

in the joint realization, which yields (4.3). Hence p_λ is increasing in λ. If

$\lambda < \lambda_*$, then $p_\lambda = \nu_{\lambda,1}$ ({0 is infected}) = 0 . Since δ_Z and ρ are translation invariant, so are the $\delta_Z P_\lambda^t$, $t \in T$, and their limit $\nu_{\lambda,1}$. Thus $\nu_{\lambda,1} = \delta_\emptyset$. In this case the system is ergodic by Theorem (3.1). If $\lambda > \lambda_*$, then $\nu_{\lambda,1}(\{0 \text{ is infected}\}) > 0$, whence $\nu_1 \neq \delta_\emptyset$. Clearly the system is nonergodic in this case. The proof for the one-sided systems is analogous. □

(4.4) Proposition. With $p_\lambda, \lambda_*, p_\lambda^+$ and λ_*^+ defined as in Proposition (4.1),

$$p_\lambda \leq \frac{2\lambda - 2}{2\lambda - 1} \qquad \lambda > 1,$$
$$= 0 \qquad \lambda \leq 1,$$

whence $\lambda_* \geq 1$, and

$$p_\lambda^+ \leq \frac{\lambda - 2}{\lambda - 1} \qquad \lambda > 2,$$
$$= 0 \qquad \lambda \leq 2,$$

whence $\lambda_*^+ \geq 2$.

Proof. We treat the basic case; the one-sided argument is similar, and will be left as an exercise. By duality,

$$p_\lambda = \hat{P}(\hat{\tau}^0_{\lambda,\emptyset} = \infty) = \hat{P}(R_{\lambda,t} - L_{\lambda,t} > 0 \quad \forall t),$$

where $\hat{\tau}^0_{\lambda,\emptyset}$ is the hitting time of $\emptyset$ for $\hat{\xi}^0_{\lambda,t}$, and $R_{\lambda,t}$ and $L_{\lambda,t}$ are defined on $\{\hat{\tau}_{\lambda,\emptyset} > t\}$ by

$$R_{\lambda,t} = \max\{x : x \in \hat{\xi}_t^0\}, \quad L_{\lambda,t} = \min\{x : x \in \hat{\xi}_t^0\}.$$

Think of $(\hat{\xi}_t^x)$ as a coalescing branching process. Whenever $D_{\lambda,t} = R_{\lambda,t} - L_{\lambda,t} > 0$, $R_{\lambda,t}$ moves one unit to the right at rate λ, and _at least_ one unit to the left at rate 1. $L_{\lambda,t}$ moves one unit left at rate λ, at least one unit to the right at rate 1. Thus D_t increases by 1 at rate 2λ, and decreases by at least 1 at rate 2, whenever $D_t \geq 1$. From value 0, D_t goes to 1 at rate 2λ, whereas the process dies out at rate 1. It follows that

$$p_\lambda \leq P_0(X_n \geq 0 \quad \forall n),$$

where (X_n) is a discrete time Markov chain on the state space $\{-1,0,1,2,\cdots\}$ with transition probabilities

$$p_{x\,x+1} = \frac{\lambda}{1+\lambda}\,, \quad p_{x\,x-1} = \frac{1}{1+\lambda} \qquad x \geq 1\,,$$

$$p_{0\,1} = \frac{2\lambda}{1+2\lambda}\,, \quad p_{0\,-1} = \frac{1}{1+2\lambda}$$

$$p_{-1\,-1} = 1\,.$$

Consider the total probability equation:

(4.5)
$$P_0(X_n \geq 0 \;\; \forall n) = \frac{2\lambda}{1+2\lambda}\,[P_1(X_n > 0 \;\; \forall n) + P_1(X_n = 0 \text{ for some } n)\,P_0(X_n \geq 0 \;\; \forall n)]\,.$$

Since X_n is a random walk when restricted to $x \geq 1$, the famous gambler's ruin formula implies that

$$P_1(X_n = 0 \text{ for some } n) = \frac{1}{\lambda} \qquad \lambda > 1$$
$$= 1 \qquad \lambda \leq 1\,.$$

Substitute in (4.5) and solve for $P_0(X_n \geq 0 \;\; \forall n)$, the desired upper bound for p_λ. □

(4.6) Problem. Derive the bounds on p_λ^+ and λ_*^+ given in Proposition (4.4).

We now turn to one of the deepest results in the theory of particle systems: the permanence of infection for contact systems with sufficiently large λ. There is no known proof that $\lambda^* < \infty$ which is truly elementary, but a remarkable method of Holley and Liggett (1978) comes the closest. We sketch their approach, referring to their paper for most of the details.

(4.7) Theorem. With p_λ, λ_*, p_λ^+ and λ_*^+ defined as in Proposition (4.1),

$$p_\lambda \geq \frac{1}{2} + \sqrt{\frac{1}{4} - \frac{1}{2\lambda}} \qquad \lambda > 2\,,$$

whence $\lambda_* \leq 2$, and

$$p_\lambda^+ \geq \frac{1}{2} + \sqrt{\frac{1}{4} - \frac{1}{\lambda}} \qquad \lambda > 4\,,$$

whence $\lambda_*^+ \leq 4$.

Sketch of proof. The basic and one-sided cases are analogous, so we discuss the former. The idea is to find a translation invariant $\mu = \mu_\lambda$ such that $\mu(\{A : 0 \in A\}) > 0$ and $\varphi^{\mu}_{\lambda,t}(\Lambda) = P(\xi^{\mu}_{\lambda,t} \cap \Lambda = \emptyset)$ is decreasing in t for all $\Lambda \in S_0$. This clearly proves nonergodicity; in fact

$$\nu_{\lambda,1}(\{0 \text{ is infected}\}) \geq 1 - \varphi^{\mu}(0) > 0 . \tag{4.8}$$

For the remainder of the discussion λ will be fixed, and often suppressed from the notation. By self-duality of the basic contact process (cf. (3.5)) and (1.10),

$$\varphi^{\mu}_t(\Lambda) = E[\varphi^{\mu}(\xi_t^{\Lambda})] \quad .$$

It therefore suffices to check that

$$\frac{d}{dt} E[\varphi^{\mu}(\xi_t^{\Lambda})] \Big|_{t=0} \leq 0 \qquad \forall \Lambda \in S_0 . \tag{4.9}$$

Unfortunately, no product measure μ_θ satisfies (4.9) for all Λ . There is, however, a renewal measure which works provided λ is large enough. The renewal measure $\mu_f \in \mathfrak{M}$ is determined by a probability density $f = (f_k)_{k=1}^{\infty}$ such that $m = \sum_{k=1}^{\infty} k f_k < \infty$. μ_f is translation invariant, with cylinder probabilities given by

$$\begin{aligned}\mu_f(\{A : A(x) = A(x+y_1) = \cdots = A(x+y_1+\cdots+y_n) = 1 ,\\ A(z) = 0 \text{ for all other } z \in [x, x+y_1+\cdots+y_n]\})\\ = m^{-1} \prod_{\ell=1}^{n} f_{y_\ell} \quad .\end{aligned}$$

The method of Holley and Liggett is to choose (f_k) so that (4.9) holds with equality in case $\Lambda = [x,y]$ for some $x \leq y$, and then to prove the inequality (4.9) for arbitrary Λ with $\mu = \mu_f$ so chosen. The all-important second part of the program is rather involved, so we will omit it, and refer the reader to Holley and Liggett (1978). To find the desired f, note that when $\Lambda = [0,n-1]$, the contact process grows one unit at either end with rate λ, while an infected site $k \in [0,n-1]$ recovers at rate 1. Thus, equality in (4.9) is equivalent to the equation

$$\sum_{k=0}^{n-1} [\varphi^{\mu}([0,n-1] - \{k\}) - \varphi^{\mu}([0,n-1])]$$

(4.10)

$$= \lambda[\varphi^{\mu}([0,n-1]) - \varphi^{\mu}([0,n])] + \lambda[\varphi^{\mu}([0,n-1]) - \varphi^{\mu}([-1,n-1])] .$$

Put $F_n = \sum_{k=n+1}^{\infty} f_k$. Then (4.10) becomes

$$2\lambda F_n = \sum_{k=0}^{n-1} F_k F_{n-k}, \quad n \geq 1 \quad (F_0 = 1). \tag{4.11}$$

To find the F_n, introduce the generating function $\Gamma(x) = \sum_{n=0}^{\infty} F_n x^n$. (4.11) is equivalent to

$$2\lambda(\Gamma(x) - 1) = x\Gamma^2(x),$$

or

$$\Gamma(x) = \frac{\lambda - \sqrt{\lambda^2 - 2\lambda x}}{x} .$$

One can solve for F to get $F_n = \frac{(2n)!}{n!(n+1)!}(2\lambda)^{-n}$, which is summable for $\lambda \geq 2$. Over this parameter range,

$$m - \sum_{k=0}^{\infty} F_k = \Gamma(1) = \lambda - \sqrt{\lambda^2 - 2\lambda}$$

Since $\mu(\{A : 0 \in A\}) = m^{-1}$, (4.8) yields the lower bound on p_λ. □

(4.12) Problem. Show that analogous computations for the one-sided systems give rise to the inequality for p_λ^+ which is stated in Theorem (4.7). (Note that both the upper and lower bounds for p_λ are precisely the same as those for $p_{2\lambda}^+$. It is an intriguing and open question as to whether $\lambda_*^+ > 2\lambda_*$, $\lambda_*^+ < 2\lambda_*$, or perhaps $\lambda_*^+ = 2\lambda_*$.)

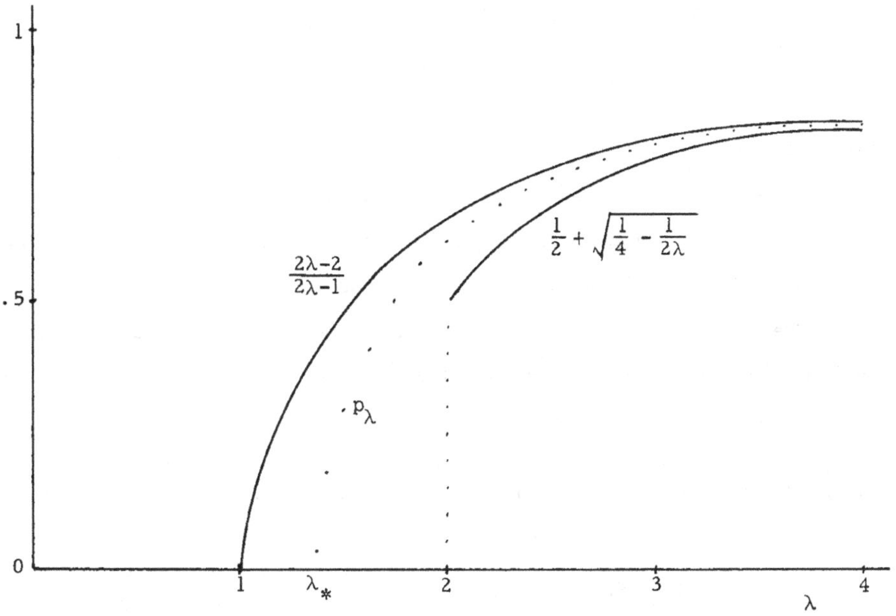

<u>figure</u> v.

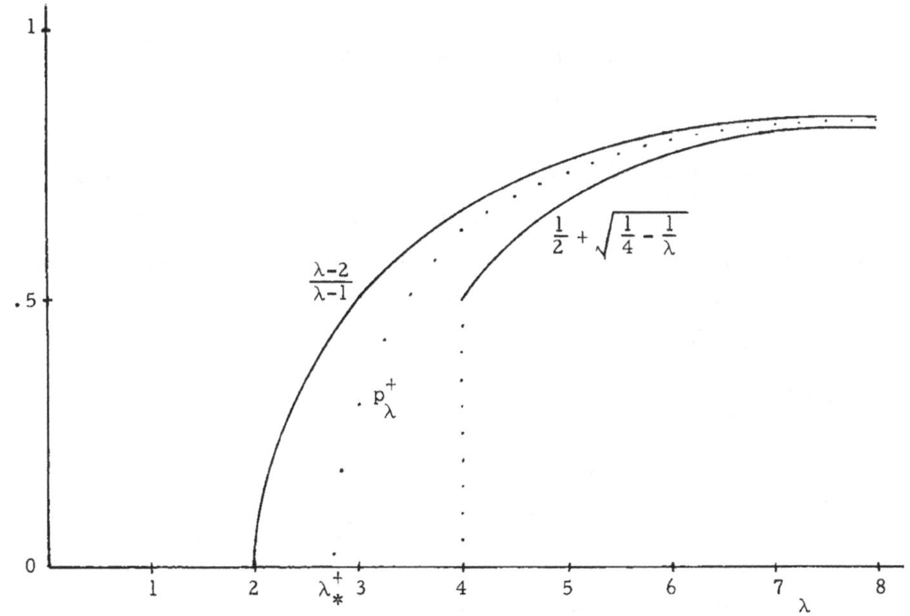

<u>figure</u> vi.

To summarize, we have seen that

$$p_\lambda = \lim_{t \to \infty} P(0 \in \xi^Z_{\lambda,t}) = P(\xi^0_{\lambda,t} \neq \emptyset \quad \forall t)$$

is increasing in λ, equals 0 for $\lambda \leq 1$, and is strictly positive for $\lambda \geq 2$. In fact, p_λ is sandwiched between the two curves shown in figure v. While we have drawn p_λ to be continuous at $\lambda = \lambda_*$, there is no known rigorous basis for this. The analogous graphs for the one-sided systems are shown in figure vi.

Theorem (4.7) gives the best known upper bounds for λ_* and λ_*^+. In contrast, there is a technique for improving the lower bounds of Proposition (4.4). We illustrate this with our next result.

(4.13) <u>Proposition.</u> Let λ_* and λ_*^+ be the critical values for the basic and one-sided contact systems respectively. Then

$$\lambda_* \geq \frac{1 + \sqrt{37}}{6} \approx 1.16, \quad \text{and} \quad \lambda_*^+ \geq \sqrt{6} \approx 2.41 \ .$$

<u>Proof.</u> We derive the first bound; the second is left as an exercise. By self-duality, it suffices to prove that

$$P(\tau^0_{\lambda,\emptyset} = \infty) = 0 \quad \text{whenever} \quad \lambda \leq \frac{1 + \sqrt{37}}{6} \ .$$

Set $\sigma(\Lambda) = P(\tau^\Lambda_{\lambda,\emptyset} = \infty)$, and note that σ is a harmonic function for the discrete time Markov chain obtained by looking at $\{(\xi^\Lambda_{\lambda,t})\}$ at its jump times $\tau_1, \tau_2, \cdots$. Also, by translation invariance, we can write $\sigma(x) \equiv \sigma(\cdot)$, $\sigma(\{x, x+1\}) \equiv \sigma(\cdot\,\cdot)$, $\sigma(\{x, x+2\}) \equiv \sigma(\cdot - \cdot)$, etc. The following total probability equations are obtained

(A) $\quad \sigma(\cdot) = \dfrac{2\lambda}{1 + 2\lambda}\, \sigma(\cdot\,\cdot)$,

(B) $\quad \sigma(\cdot\,\cdot) = \dfrac{1}{1 + \lambda}\, \sigma(\cdot) + \dfrac{\lambda}{1 + \lambda}\, \sigma(\cdot\,\cdot\,\cdot)$,

(C) $\quad \sigma(\cdot - \cdot) = \dfrac{1}{1 + 2\lambda}\, \sigma(\cdot) + \dfrac{\lambda}{1 + 2\lambda}\, \sigma(\cdot\,\cdot\,\cdot) + \dfrac{\lambda}{1 + 2\lambda}\, \sigma(\cdot\,\cdot - \cdot)$,

(D) $\quad \sigma(\cdot\,\cdot\,\cdot) = \dfrac{2}{3 + 2\lambda}\, \sigma(\cdot\,\cdot) + \dfrac{1}{3 + 2\lambda}\, \sigma(\cdot - \cdot) + \dfrac{2\lambda}{3 + 2\lambda}\, \sigma(\cdot\,\cdot\,\cdot\,\cdot)$.

Observe next that σ is strongly subadditive:

$$\sigma(\Lambda_1 \cup \Lambda_2) \le \sigma(\Lambda_1) + \sigma(\Lambda_2) - \sigma(\Lambda_1 \cap \Lambda_2) \qquad \Lambda_1, \Lambda_2 \in S_0 \ .$$

In fact, by duality, $\sigma(\Lambda_1 \cup \Lambda_2) - \sigma(\Lambda_1) = \varphi^{\nu_1}(\Lambda_1) - \varphi^{\nu_1}(\Lambda_1 \cup \Lambda_2)$
$= \nu_1(\{A : A \cap \Lambda_1 = \emptyset,\ A \cap (\Lambda_1^c \cap \Lambda_2) \ne \emptyset\}) \le \nu_1(\{A : A \cap (\Lambda_1 \cap \Lambda_2) = \emptyset,\ A \cap (\Lambda_1^c \cap \Lambda_2) \ne \emptyset\})$
$= \varphi^{\nu_1}(\Lambda_1 \cap \Lambda_2) - \varphi^{\nu_1}(\Lambda_2) = \sigma(\Lambda_2) - \sigma(\Lambda_1 \cap \Lambda_2)$. In particular,
$\sigma(\cdot\cdot - \cdot) \le \sigma(\cdot\cdot) + \sigma(\cdot - \cdot) - \sigma(\cdot)$ and $\sigma(\cdot\cdot\cdot\cdot) \le 2\sigma(\cdot\cdot\cdot) - \sigma(\cdot\cdot)$.
Substituting into (C) and (D) we get the new inequalities

$$\text{(C')} \qquad \sigma(\cdot - \cdot) \le \frac{1-\lambda}{1+\lambda}\,\sigma(\cdot) + \frac{\lambda}{1+\lambda}\,\sigma(\cdot\cdot) + \frac{\lambda}{1+\lambda}\,\sigma(\cdot\cdot\cdot) ,$$

$$\text{(D')} \qquad \sigma(\cdot\cdot\cdot) \le \frac{2(1-\lambda)}{3-2\lambda}\,\sigma(\cdot\cdot) + \frac{1}{3-2\lambda}\,\sigma(\cdot - \cdot) \ .$$

Now if $0 \le \lambda < \frac{1+\sqrt{37}}{6}$ (= the positive root of $3 + \lambda - 3\lambda^2 = 0$), the positive combination:

$$(\lambda+2)(3+\lambda-3\lambda^2)(A) + 2\lambda(1+\lambda)(3-2\lambda)(B)$$

$$+ 2\lambda^2(1+\lambda)(C') + 2\lambda^2(3-2\lambda)(1+\lambda)(D')$$

yields

$$(\lambda+2)(3+\lambda-3\lambda^2)\,\sigma(\cdot) \le 2\lambda(3+\lambda-3\lambda^2)\,\sigma(\cdot) ,$$

so that $\sigma(\cdot)$ must equal 0 . This completes the proof. We remark that better bounds can be obtained if one is willing to handle larger systems of inequalities. □

(4.14) Problem. Derive the bound $\lambda_*^+ \ge \sqrt{6}$ by applying the same method to the one-sided systems.

(4.15) Notes. Contact processes were first studied systematically by Harris (1974), although work on closely related systems, especially in discrete time, had been carried out by Soviet probabilists for several years. See especially Dobrushin (1971), Stavskaya and Pyatetakii-Shapiro (1968), Toom (1968), Vasilev (1969) and Vasilev et al. (1973). Versions of Proposition (4.4) may be found in Harris (1974) and Holley and Liggett (1975). Permanence of contact systems for large λ was proved by Harris (1974); his method was based on comparison with discrete time systems and appeal to the percolation techniques of

Hammersley (1959) and Toom (1968). The computations for Proposition (4.13) are taken from Griffeath (1975).

5. Contact systems: limit theorems in the nonergodic case.

In this section we study the limiting behavior of nonergodic contact systems. $\{(\xi_t^A)\}$ will be the basic system, $\{(\xi_t^{+,A})\}$ the one-sided system, with λ prescribed. Our first result is a "complete convergence theorem" for $\{(\xi_t^A)\}$; unfortunately the method of proof only works for $\lambda > \lambda_*^+$.

(5.1) Theorem. Let $\{(\xi_t^A)\}$ be the basic (one-dimensional) contact system with infection parameter λ. If $\lambda > \lambda_*^+$ (= the critical value for the one-sided system), then for any $\mu \in \mathfrak{M}$,

$$\mu P^t \to P(\tau_\emptyset^\mu < \infty)\,\delta_\emptyset + P(\tau_\emptyset^\mu = \infty)\,\nu_1 \qquad \text{as} \quad t \to \infty ,$$

where $\tau_\emptyset^\mu = \inf\{t \in T : \xi_t^\mu = \emptyset\}$ (= ∞ if no such t exists).
In particular, $\delta_\emptyset$ and ν_1 are the only extreme invariant measures for $\{(\xi_t^A)\}$.

Proof. It will be convenient to establish some preliminary results in the form of two lemmas.

(5.2) Lemma. If $\{(\xi_t^A)\}$ is nonergodic, so that $\nu_1 \neq \delta_\emptyset$, then $\nu_1(S_0) = 0$; also

$$(5.3) \qquad \lim_{N\to\infty} \ \sup_{\Lambda : |\Lambda| = N} \ \varphi^{\nu_1}(\Lambda) = 0 ,$$

so that

$$(5.4) \qquad \lim_{N\to\infty} \ \sup_{\Lambda : |\Lambda| = N} \ P(\tau_\emptyset^\Lambda < \infty) = 0 .$$

The same results hold for $\{(\xi_t^{+,A})\}$.

Sketch of proof. If $\Lambda \in S_0$, $\Lambda \neq \emptyset$, then $\nu_1(\{\Lambda\})$ cannot be positive, since all the translates of Λ have the same ν_1-probability. To show $\nu_1(S_0) = 0$, it therefore suffices to check that $c = \nu_1(\{\emptyset\}) = 0$. Since $(\xi_t^{\nu_1})$ is stationary,

$$\varphi^{\nu_1}([-n,n]) = c + \int_{A \neq \emptyset} \varphi_t^A([-n,n])\nu_1(dA)$$
$$\geq c + \int_{A \neq \emptyset} \varphi_t^Z([-n,n])\nu_1(dA)$$
$$= c + (1-c)\varphi_t^Z([-n,n]) \ .$$

Let $t \to \infty$ to get $c\varphi^{\nu_1}([-n,n]) \geq c$. Now let $n \to \infty$ to force $c = 0$. The proof of (5.3) is somewhat technical, so we omit it, and refer the reader to Harris (1974). Property (5.4) is equivalent to (5.3) by self-duality. Very similar arguments apply to the one-sided process. □

(5.5) Lemma. For $\Lambda \in S_0$, $t < \tau_\emptyset^\Lambda$, define

$$L_t^\Lambda = \min\{x : x \in \xi_t^\Lambda\}, \quad R_t^\Lambda = \max\{x : x \in \xi_t^\Lambda\} \ .$$

For any $x \in Z$, $t \in T$,

$$\text{(5.6)} \qquad \xi_t^x = \xi_t^Z \cap [L_t^x, R_t^x] \quad \text{on} \quad \{t < \tau_\emptyset^x\} \ .$$

If $\lambda > \lambda_*^+$, then for any $\Lambda \in S_0$,

$$\text{(5.7)} \qquad L_t^\Lambda \to -\infty \quad \text{and} \quad R_t^\Lambda \to \infty \qquad P-\text{a.s.} \quad \text{on} \quad \{\tau_\emptyset^\Lambda = \infty\} \ .$$

Proof. Since $\xi_t^x \subset \xi_t^Z \cap [L_t^x, R_t^x]$ by monotonicity and the definitions of L_t^x and R_t^x, it remains to check the opposite inclusion. Suppose $z \in \xi_t^Z \cap [L_t^x, R_t^x]$. Then $N_t^y(z) > 0$ for some $y \in Z^d$. If $y = x$, then $z \in \xi_t^x$. If $y < x$, then a path up from $(y,0)$ to (z,t) intersects a path up from $(x,0)$ to (L_t^x, t). By following the latter path up to the intersection point, and then following the former, we get a path up from $(x,0)$ to (z,t). Hence $z \in \xi_t^x$. The same argument applies if $y > x$ by using R_t^x instead. This completes the proof of (5.6). We remark that the nearest neighbor nature of the infection mechanism is crucial in constructing the composite path, for otherwise paths can "jump over one another." Turning to the proof of (5.7), we introduce

$$\tau_N^\Lambda = \min\{t \in T : |\xi_t^\Lambda| = N\} \quad (= \infty \text{ if no such } t \text{ exists}), \quad N \geq 1 \ .$$

For $N \geq |\Lambda|$, since $\emptyset$ is a trap, a standard Markov process argument shows that $\tau_N^\Lambda < \infty$ P-a.s. on $\{\tau_\emptyset^\Lambda = \infty\}$. Hence

$$P(\liminf_{t\to\infty} R_t^\Lambda < \infty\,,\ \tau_\emptyset^\Lambda = \infty)$$

$$= \int_{s\in(0,\infty)} \sum_{B:|B|=N} P(\tau_N^\Lambda \in ds,\ \xi_{\tau_N^\Lambda}^\Lambda = B)\, P(\liminf_{t\to\infty} R_t^B < \infty\,,\ \tau_\emptyset^B = \infty)\ .$$

To get the result for R_t^Λ, we check that

$$(5.8) \qquad \lim_{N\to\infty} \sup_{B:|B|=N} P(\liminf_{t\to\infty} R_t^B < \infty,\ \tau_\emptyset^B = \infty) = 0\ .$$

Now $R_t^B \geq R_t^{+,B}$ = the right hand edge of the one-sided process defined on the same substructure by using only positive-oriented arrows, on $\{\tau_\emptyset^{+,B} = \infty\} \subset \{\tau_\emptyset^B = \infty\}$. Hence we need only check condition (5.8) for the one-sided system. For those processes $R_t^{+,B} \to +\infty$ P - a.s. on $\{\tau_\emptyset^{+,B} = \infty\}$, so an application of (5.4) (for $\{(\xi_t^{+,A})\}$) completes the proof. □

<u>Proof of Theorem</u> (5.1). The first step is to show that for each $B \in S - \{\emptyset\}$, $\Lambda \in S_0$,

$$(5.9) \qquad P(\xi_t^B \cap \Lambda = \xi_t^Z \cap \Lambda \text{ for all large } t \mid \tau_\emptyset^B = \infty) = 1\ .$$

Note that if $\tau_\emptyset^B = \infty$, then $\tau_\emptyset^x = \infty$ for some $x \in B$. If $B \in S_0$ this follows from additivity, while for $B \in S_\infty$ we have
$P(\tau_\emptyset^B = \infty) = \lim_{N\to\infty} P(\tau_\emptyset^{B\cap[-N,N]} = \infty) = 1$ by (5.4). Thus if $\tau_\emptyset^B = \infty$, then choosing $x \in B$ so that $\tau_\emptyset^x = \infty$, we see from (5.6) that $\xi_t^B \cap \Lambda = \xi_t^Z \cap \Lambda$ as soon as $\Lambda \subset [R_t^x, L_t^x]$. Applying (5.7), we obtain (5.9). Next, write

$$(5.10) \qquad \begin{aligned} \varphi_t^B(\Lambda) &= P(\tau_\emptyset^B \leq t) + P(\xi_t^B \cap \Lambda = \emptyset,\ t < \tau_\emptyset^B < \infty) \\ &\quad + P(\xi_t^B \cap \Lambda = \emptyset \mid \tau_\emptyset^B = \infty)\, P(\tau_\emptyset^B = \infty) \end{aligned}$$

Suppose we can show

$$(5.11) \qquad \lim_{t\to\infty} P(\xi_t^Z \cap \Lambda = \emptyset \mid \tau_\emptyset^B = \infty) = \varphi^{\nu_1}(\Lambda)\ .$$

Then letting $t\to\infty$ in (5.10), and using (5.9),

$$\lim_{t\to\infty} \varphi_t^B(\Lambda) = P(\tau_\emptyset^B < \infty)\varphi^\emptyset(\Lambda) + P(\tau_\emptyset^B = \infty)\varphi^{\nu_1}(\Lambda)\ .$$

The desired convergence theorem follows by inclusion-exclusion. It therefore suffices to check (5.11), or equivalently

$$(5.12)\qquad \lim_{t\to\infty} P(\xi_t^Z \cap \Lambda = \emptyset,\ \tau_\emptyset^B > t) = P(\tau_\emptyset = \infty)\varphi^{\nu_1}(\Lambda)\ .$$

For $s < t$,

$$P(\xi_t^Z \cap \Lambda = \emptyset,\ \tau_\emptyset^B > s) = P(\tau_\emptyset^B > s)\sum_C P(\xi_s^Z = C \mid \tau_\emptyset^B > s)\varphi_{t-s}^C(\Lambda)$$
$$\geq P(\tau_\emptyset^B > s)\varphi_{t-s}^Z(\Lambda)\ .$$

So

$$\liminf_{t\to\infty} P(\xi_t^Z \cap \Lambda = \emptyset,\ \tau_\emptyset^B > s) \geq P(\tau_\emptyset^B > s)\varphi^{\nu_1}(\Lambda)\ .$$

Similarly,

$$\liminf_{t\to\infty} P(\xi_t^Z \cap \Lambda = \emptyset,\ \tau_\emptyset^B \leq s) \geq P(\tau_\emptyset^B \leq s)\varphi^{\nu_1}(\Lambda)\ .$$

But we know that $\lim_{t\to\infty} \varphi_t^Z(\Lambda) = \varphi^{\nu_1}(\Lambda)$, so

$$\lim_{t\to\infty} P(\xi_t^Z \cap \Lambda = \emptyset,\ \tau_\emptyset^B > s) = P(\tau_\emptyset^B > s)\varphi^{\nu_1}(\Lambda)\ .$$

Let $L = \lim_{s\to\infty} \lim_{t\to\infty} P(\xi_t^Z \cap \Lambda = \emptyset,\ \tau_\emptyset^B > s)$. Then $L = P(\tau_\emptyset^B = \infty)\varphi^{\nu_1}(\Lambda)$, so to get (5.12) and finish the proof we need $L = \lim_{t\to\infty} P(\xi_t^Z \cap \Lambda = \emptyset,\ \tau^B > t)$. But this is clear, since

$$0 \leq \lim_{s\to\infty} \lim_{t\to\infty} [P(\xi_t^Z \cap \Lambda = \emptyset,\ \tau_\emptyset^B > s) - P(\xi_t^Z \cap \Lambda = \emptyset,\ \tau_\emptyset^B > t)]$$
$$\leq \lim_{s\to\infty} P(s < \tau_\emptyset^B < \infty) = 0\ . \qquad \square$$

Using some of the same observations one can prove a corresponding "complete pointwise ergodic theorem" for $\{(\xi_t^A)\}$.

(5.13) <u>Theorem</u>. Let $\{(\xi_t^A)\}$ be the basic contact system $(d = 1)$ with parameter λ. If $\lambda > \lambda_*^+$, then for any $\mu \in \mathfrak{M}$, $f \in \mathcal{C}$,

$$\frac{1}{t}\int_0^t f(\xi_s^\mu)\,ds \to f(\emptyset) \text{ as } t\to\infty \quad P\text{-a.s. on } \{\tau_\emptyset^\mu < \infty\},$$
$$\to \int_S f\,d\nu \text{ as } t\to\infty \quad P\text{-a.s. on } \{\tau_\emptyset^\mu = \infty\}\ .$$

Proof. The first assertion is obvious. To prove the second, we proceed as for Theorem (2.9). Problem (I.1.10) applies to ν_1, by Theorem (5.1) and Lemma (5.2). Thus the stationary process $(\xi_t^{\nu_1})$ is Birkhoff ergodic, and so

$$P\left(\frac{1}{t}\int_0^t f(\xi_s^{\nu_1})\,ds \to \int_S f\,d\nu_1\right) = 1 \qquad \forall f \in L^1(\nu_1)\ .$$

Fix $B \in S$, $f \in \mathfrak{I} = \{\text{cylinder functions}\}$. Property (5.9), the fact that $P(\tau_\emptyset^A = \infty) = 1$ for any $A \in S_\infty$, and $\nu(S_\infty) = 1$ together imply

$$P(f(\xi_s^B) = f(\xi_s^{\nu_1}) \text{ for all large } s \mid \tau_\emptyset^B = \infty) = 1\ .$$

Defining $\tau_f = \min\{t : f(\xi_s^B) = f(\xi_s^{\nu_1})\}$, $\tau_f < \infty$ P - a.s. on $\{\tau_\emptyset^B = \infty\}$. Thus, on $\{\tau_\emptyset^B = \infty\}$, we can argue just as in the proof of Theorem (2.9) to get the desired convergence when $\mu = \delta_B$, $f \in \mathfrak{I}$. The extensions to general μ and $f \in \mathcal{C}$ are straightforward. □

Next, we mention an application of the graphical representation which is based on the same observation that was used in the proof of (5.6). Harris (1978) proves that the basic contact system $\{(\xi_t^A)\}$ satisfies a more general collection of correlation inequalities than those in Theorem (2.17), namely:

(5.14) $$P(\xi_t^A \cap C = \emptyset,\ \xi_t^B \cap D = \emptyset) \geq \varphi_t^A(C)\,\varphi_t^B(D)$$

$A, B, C, D \in S$, $t \in T$. He then goes on to show that (5.14) has a simple but beautiful consequence.

(5.15) Theorem. Let $\{(\xi_t^A)\}$ be a nonergodic basic contact system $(d = 1)$. Then

$$\liminf_{t \to \infty} P(0 \in \xi_t^0) > 0\ .$$

Proof. If the system is nonergodic, then $\varepsilon = \nu_1(0 \text{ is infected}) = P(\tau_\emptyset^0 = \infty) > 0$, and

$$P(0 \in \xi_t^Z) = P(\tau_\emptyset^0 > t) \geq \varepsilon \qquad \forall t\ .$$

Let $Z^+ = \{0, 1, 2, \cdots\}$. By symmetry

$$P(0 \in \xi_t^{Z^+}) = P(\xi_t^0 \cap Z^+ \neq \emptyset) \geq \frac{\varepsilon}{2} \qquad \forall t\ .$$

Positive correlation of $\{\xi_t^A \cap C \neq \emptyset\}$ and $\{\xi_t^B \cap D \neq \emptyset\}$ is equivalent to (5.14); taking $A = D = Z^+$, $C = B = \{0\}$, we get

$$P(0 \in \xi_t^{Z^+}, \xi_t^0 \cap Z^+ \neq \emptyset) \geq \frac{\varepsilon^2}{4} \qquad \forall t .$$

Since paths cannot jump over one another, $N_t^{Z^+}(0) > 0$ and $N_t^0(Z^+) > 0$ together imply $N_t^0(0) > 0$. Hence

$$\liminf_{t \to \infty} P(0 \in \xi_t^0) \geq \frac{\varepsilon^2}{4} ,$$

as desired. We remark that for $\lambda > \lambda_*^+$, Theorem (5.1) yields $\liminf_{t \to \infty} P(0 \in \xi_t^0) \geq \varepsilon^2$. The present result is of interest because it applies all the way down to the critical constant. □

We conclude this section with a brief survey of further results for the basic and one-sided one-dimensional contact systems. First, we note that <u>none</u> of the last three theorems hold for the $\{(\xi_{\lambda,t}^{+,A})\}$. In fact, if $A \in S_0$ then $\delta_A P^t \to \delta_\emptyset$ for <u>any</u> λ: in the nonergodic case the set of infected sites wanders off to the right if it does not die out. By taking A to be a countable disjoint union of larger and larger blocks which are farther and farther apart, and by taking $\lambda > \lambda_*^+$, one can get examples where $\delta_A P^t$ does not converge as $t \to \infty$, but rather converges to ν_1 along one subsequence and to $\delta_\emptyset$ along another.

Liggett (1978) has shown that both the basic and one-sided systems have only $\delta_\emptyset$ and ν_1 as extreme invariant measures for <u>all</u> parameter values λ. Also, convergence to ν_1 from "nice" initial measures μ takes place in both systems for all parameter values, as a special case of a result to be mentioned in the next section. The questions of convergence and pointwise ergodicity for the basic systems with λ just above λ_*, and starting from arbitrary $\mu \in \mathfrak{M}$, remain open.

(5.16) <u>Notes</u>. Theorem (5.1) is proved in Griffeath (1978a). A similar discrete time result was obtained by Vasilev (1969), using the contour method of percolation theory. The proof of (5.1) which we give here is due to Liggett (private communication). There is a sketch of this proof in the introduction of Liggett (1978); it has the advantage of leading to (a) Liggett's theorem that $\mathcal{J}$ is one-dimensional in the nonergodic case, and (b) Theorem (5.13) (which is taken from Griffeath (1979)).

6. Contact systems in several dimensions.

There are many ways to generalize the basic contagion model studied in the last section. We consider here only the most natural generalization to Z^d, $d \geq 1$. By the basic d-dimensional contact system we mean the lineal proximity system on Z^d with $I_x \equiv \{0,1,\cdots,2d\}$, $\lambda_{0,x} \equiv 1$, $C_{0,x} \equiv \emptyset$, $\lambda_{i,x} \equiv \lambda$ for $i \geq 1$, and $C_{i,x} = \{x,y_i\}$ for $i \geq 1$, where the y_i are the $2d$ sites immediately adjacent to site x. In words, infected particles recover at rate 1, while infection takes place at a rate proportional to the number of infected neighbors. (x and y in Z^d are neighbors if $|x-y| = 1$.) The proportionality constant for the rate of infection is given by the parameter λ.

Less is known about several-dimensional contact systems, but the existence of a critical λ_*^d in each dimension d can be proved just as in the one-dimensional case. In fact, using some of the methods already discussed, one obtains the following results.

(6.1) Theorem. Let $\{(\xi_{\lambda,t}^A)\}$ be the basic d-dimensional contact system with parameter λ, and set $p_\lambda = \nu_1(0 \text{ is infected})$. Then p_λ is increasing in λ. If $\lambda_*^d = \sup\{\lambda : p_\lambda = 0\}$, then the system is ergodic for $\lambda < \lambda_*^d$, nonergodic for $\lambda > \lambda_*^d$, and

(6.2) $$\frac{1}{2d-1} \leq \lambda_*^d \leq \frac{2}{d} .$$

Proof. The argument for everything except (6.2) is very similar to the proof of Proposition (4.1), so we omit it. To get the left hand inequality in (6.2), apply the method of Proposition (4.13). Namely, since $\{(\xi_{\lambda,t}^A)\}$ is self-dual, it suffices to prove that $P(\tau_{\lambda,\emptyset}^0 = \infty) = 0$ whenever $\lambda < \frac{1}{2d-1}$. Let $\sigma(\Lambda) = P(\tau_{\lambda,\emptyset}^\Lambda = \infty)$, and note that σ satisfies the total probability equations

(A) $$\sigma(\{0\}) = \frac{2d\lambda}{1+2d\lambda}\,\sigma(\{0,e_1\}) ,$$

(B) $$\sigma(\{0,e_1\}) = \frac{1}{1+(2d-1)\lambda}\,\sigma(\{0\}) + \frac{\lambda}{1+(2d-1)\lambda} \sum_{j=2}^{2d} \sigma(\{0,e_1,e_2\}) .$$

(Here we have made use of the translation invariance of σ .) Also, by strong subadditivity, $\sigma(\{0,e_1,e_j\}) \le 2\sigma(\{0,e_1\}) - \sigma(\{0\})$, so from (B) we get

$$(1-(2d-1)\lambda)\,\sigma(\{0,e_1\}) \le (1-(2d-1)\lambda)\,\sigma(\{0\}) \ .$$

If $\lambda < \frac{1}{2d-1}$, then $\sigma(\{0,e_1\}) \le \sigma(\{0\})$. Substituting into (A), we find that

$$\sigma(\{0\}) \le \frac{2d\lambda}{1+2d\lambda}\,\sigma(\{0\}) ,$$

which implies $\sigma(\{0\}) = 0$, the desired result. The right hand bound in (6.2) is obtained by comparison with the one-dimensional systems. Namely, define a stochastic process (ζ_t^0) on Z by

$$\zeta_t^0 = \{x \in Z : \sum_{k=1}^{d} x_k = x \text{ for some } (x_1,\cdots,x_d) \in \xi_{\lambda,t}^0\} \ .$$

One can check that if $\zeta_t^0 = A$, then a 1 flips to a 0 with rate at most 1 , while a 0 at x flips to a 1 with rate at least $d\lambda\,|A \cap \{x-1,x+1\}|$. (A picture of the case d = 2 will help make this clear.) Thus the contagion in (ζ_t^0) is stronger than that in the basic one-dimensional system with parameter $d\lambda$. Since $\{\zeta_t^0 = \emptyset\} = \{\xi_{\lambda,t}^0 = \emptyset\}$, Theorem (4.7) yields

$$P(\tau_{\lambda,\emptyset}^0 = \infty) \ge \frac{1}{2} + \sqrt{\frac{1}{4} - \frac{1}{2d\lambda}} \qquad \lambda \ge \frac{2}{d} \ .$$

In particular, $\lambda_*^d \le \frac{2}{d}$. □

(6.3) Problem. By pushing the "strong subadditivity method" farther, show that $\lambda_*^2 \ge .359$.

Virtually all of the known dimension-independent results for nonergodic contact systems are due to Harris (1976, 1978); unfortunately his methods require regularity assumptions on the initial state. Say that $A \in S_\infty$ is n-dense $(n = 0,1,\cdots)$ if $A \cap b_n(x) \ne \emptyset$ for all $x \in Z^d$ ($b_n(x)$ as in (I.1)). A is dense if it is n-dense for some n . A measure $\mu \in \mathfrak{M}$ is called regular if

$$\lim_{n\to\infty} \sup_{x \in Z^d} [\varphi^\mu(b_n(x)) - \mu(\{\emptyset\})] = 0 \ .$$

Note that δ_A is regular if A is dense, and that any translation invariant μ is regular. The convergence theorem of Harris states that for any parameter value λ, if $\{(\xi_t^A)\}$ is a basic d-dimensional contact system (or any of a large class of contact systems which includes the one-sided system on Z), and if μ is regular, then

$$\mu P^t \to \mu(\{\emptyset\})\,\delta_\emptyset + (1-\mu(\{\emptyset\}))\nu_1 \qquad \text{as } t\to\infty .$$

This implies that the only translation invariant equilibria are mixtures of $\delta_\emptyset$ and ν_1, but it is not known whether there are additional nontranslation invariant equilibria when $d \geq 2$. Pointwise ergodic convergence to ν_1 has been proved in dimension $d \geq 2$ only for initial measures δ_A with A dense, and then only for very large λ. Finally, there is a growth rate theorem: for basic contact systems on Z^d with sufficiently large infection rates λ:

$$P(\liminf_{t\to\infty} \frac{|\xi^0_{\lambda,t}|}{t} > 0 \mid \tau^0_\emptyset = \infty) = 1 .$$

One can show that the growth is of order at most t^d, but the exact order remains open for $d \geq 2$. This concludes our discussion of contact processes.

(6.4) Notes. The lower bound in (6.2) is due to Harris (1974) and Holley and Liggett (1975). It improves a result of Dobrushin (1971). The upper bound in (6.2) is from Holley and Liggett (1978). Problem (6.3) is based on a computation in Griffeath (1975). The rest of the results mentioned in this section may be found in Harris (1974, 1976, 1978), except for some refinements in Griffeath (1978). Similar techniques were applied to discrete time systems by Vasershtein and Leontovich (1970).

7. Voter models.

This section is devoted to the study of lineal proximity systems with the property that $|C_{i,x}| = 1$ for all $i \in I_x$, the so-called voter models. According to (2.12), such a system has flip rates which can be written in the form

$$c_x(A) = \lambda_x A(x) + (1-2A(x)) \sum_{z\in A} \lambda_{z,x}$$

($\lambda_{z,x} \geq 0$). To simplify matters, we will treat only translation invariant voter models, whose flip rates can be written in the normalized form

$$c_x(A) = \lambda [A(x) + (1-2A(x)) \sum_{z \in A} p_{z-x}] \quad , \tag{7.1}$$

for some probability density $p = (p_z ; z \in Z^d)$ and $\lambda > 0$. To explain the name "voter model," we think of the sites of Z^d as occupied by persons who are either in favor of or opposed to some proposition (say 1 = "for", 0 = "against"). The "voter" at x is influenced by voter y with weight p_{y-x}, and changes opinion at a rate proportional to the sum of weights of voters with the opposite opinion. In particular, the "total consensus" states $\delta_\emptyset$ and δ_{Z^d} are <u>both traps</u> for the system. Since we are interested in asymptotic behavior of the model, and the factor λ may be removed by a change of time scale, we will assume henceforth that $\lambda = 1$. The foremost question for the voter models is: Starting from a state of "individual independence," i.e. a product measure μ_θ, does the interaction lead to "eventual unanimity" or not?

The dual systems for voter models are coalescing branching systems in which each branching tries to replace a particle by another single particle. In other words, these are <u>coalescing random walks</u>. Particles attempt to execute independent continuous time random walks with mean-1 exponential holding times and transition density p, but coalesce upon collision. In particular, the one-particle dual process is merely a random walk with density p. We say that p is <u>recurrent</u> or <u>transient</u> according to which property this random walk enjoys. The <u>basic d-dimensional voter model</u> is the system such that

$$p_z = \frac{1}{2d} \qquad |z| = 1$$
$$= 0 \qquad \text{otherwise} ,$$

i.e. the voter model whose one particle dual is simple d-dimensional random walk. Given a density p, define the symmetrization $\bar{p}$ of p by $\bar{p}_z = \frac{p_z + p_{-z}}{2}$. Note that if (X_t^1) and (X_t^2) are independent continuous time walks with density p, then $(X_t^2 - X_t^1)$ is a random walk with density $\bar{p}$ (and mean $\frac{1}{2}$ exponential holding times.) The fundamental result for voter models is that eventual unanimity occurs if $\bar{p}$ is recurrent, but disagreement persists if $\bar{p}$ is transient. Thus the basic voter model behaves one way in dimensions one and two, but entirely differently in dimension three or more. These assertions are made precise as follows.

(7.2) Theorem. Let $\{(\xi_t^A)\}$ be the (translation invariant) voter model on Z^d with flip rates (7.1) for some irreducible density p. Let $\bar{p}$ be the symmetrization of p. If $\bar{p}$ is recurrent, then

$$\mu P^t \to (1-\theta)\,\delta_\emptyset + \theta\,\delta_{Z^d} \qquad \text{as} \quad t \to \infty \tag{7.3}$$

for any initial measure μ such that $\varphi^\mu(\{z\}) \equiv 1-\theta$. If $\bar{p}$ is transient, then corresponding to each $\theta \in (0,1)$ there is a distinct translation invariant equilibrium ν_θ, with $\varphi^{\nu_\theta}(\{z\}) \equiv 1-\theta$ but not a mixture of $\delta_\emptyset$ and δ_{Z^d}, such that

$$\mu_\theta P^t \to \nu_\theta \qquad \text{as} \quad t \to \infty . \tag{7.4}$$

Moreover, each ν_θ is mixing with respect to translations in Z^d.

Proof. If $\varphi^\mu(\{z\}) = 1-\theta$ for all $z \in Z^d$, then by (1.10), $\varphi_t^\mu(\{x\}) = \hat{E}[\varphi^\mu(\hat{\xi}_t^x)] = 1-\theta$ for all $t \in T$, since $\hat{\xi}_t^x = \{z\}$ for some z. (The one particle dual is a random walk.) To prove (7.3) it suffices to show that

$$\lim_{t\to\infty} P(\xi_t^\mu(x) \neq \xi_t^\mu(y)) = 0 \qquad \forall x,y \in Z^d .$$

Equivalently we check that

$$\lim_{t\to\infty} \varphi_t^\mu(\{x,y\}) = 1-\theta \qquad \forall x,y \in Z^d . \tag{7.5}$$

A key fact about coalescing random walks is that $\hat{N}_t^A = |\hat{\xi}_t^A|$ is nonincreasing as $t \to \infty$, and always at least 1 if $A \neq \emptyset$, so that $\hat{N}_\infty^A = \lim_{t\to\infty} |\hat{\xi}_t^A|$ exists $\hat{P}$-a.s. Thus

$$\begin{aligned} \varphi_t^\mu(\{x,y\}) &= \hat{E}[\varphi^\mu(\hat{\xi}_t^{\{x,y\}})] \\ &= (1-\theta)\,\hat{P}(\hat{N}_t^{\{x,y\}} = 1) + \hat{E}[\varphi^\mu(\hat{\xi}_t^{\{x,y\}}),\ \hat{N}_t^{\{x,y\}} = 2] . \end{aligned} \tag{7.6}$$

Assume $\bar{p}$ recurrent. Since $(\xi_t^{\{x,y\}})$ acts as two independent random walks with density p until a collision occurs, we have $\hat{P}(\hat{N}_\infty^{\{x,y\}} = 1) = 1$ for all $x,y \in Z^d$. Let $t \to \infty$ in (7.6) to get (7.5). Next, assume $\bar{p}$ transient. Letting $t \to \infty$

in (1.11),

$$\lim_{t\to\infty} \varphi_t^{\mu_\theta}(\Lambda) = \hat{E}[(1-\theta)^{\hat{N}_\infty^\Lambda}] ,$$

so $\mu_\theta P^t$ converges to a measure ν_θ such that

$$\varphi^{\nu_\theta}(\Lambda) = \hat{E}[(1-\theta)^{\hat{N}_\infty^\Lambda}] . \tag{7.7}$$

Clearly ν_θ is a translation invariant equilibrium such that $\varphi^{\nu_\theta}(\{z\}) \equiv 1-\theta$. To see that ν_θ is not a mixture of $\delta_\emptyset$ and δ_{Z^d} we need only check that (7.5) does not hold for $\mu = \mu_\theta$. But for $x \neq y$,

$$\varphi^{\nu_\theta}(\{x,y\}) = (1-\theta)\hat{P}(\hat{N}_\infty^{\{x,y\}} = 1) + (1-\theta)^2\hat{P}(\hat{N}_\infty^{\{x,y\}} = 2)$$

$$= (1-\theta) - \theta(1-\theta)\hat{P}(\hat{N}_\infty^{\{x,y\}} = 2) \neq 1-\theta$$

provided $\theta \in (0,1)$ and $\hat{P}(\hat{N}_\infty^{\{x,y\}} = 2) > 0$. This last probability is positive since $\bar{p}$ is transient. To finish the proof, it remains only to show that ν_θ is spatially mixing, i.e. for each $B, C \in S_0 - \{\emptyset\}$,

$$\lim_{|z|\to\infty} [\varphi^{\nu_\theta}(B \cup (z+C)) - \varphi^{\nu_\theta}(B)\varphi^{\nu_\theta}(C)] = 0 . \tag{7.8}$$

By duality, the quantity in brackets equals

$$\hat{E}[(1-\theta)^{\hat{N}_\infty^{B\cup(z+C)}}] - \hat{E}[(1-\theta)^{\hat{N}_\infty^B}]\,\hat{E}[(1-\theta)^{N_\infty^C}] .$$

Recall from Theorem (2.17) that $(\hat{\xi}_t^{B\cup C})$ can be constructed from <u>independent</u> copies of $(\hat{\xi}_t^B)$ and $(\hat{\xi}_t^C)$ in such a way that

$$\hat{\xi}_t^{B\cup C} = \hat{\xi}_t^B \cup \hat{\xi}_t^C \quad \text{if} \quad \hat{\xi}_s^B \cap \hat{\xi}_s^C = \emptyset \qquad \forall s \le t .$$

For the remainder of the proof we will be referring to that construction. Thus we can assume that $\hat{N}_\infty^{B\cup C} = \hat{N}_\infty^B + \hat{N}_\infty^C$ if the two independent processes never interact. Hence

$$|\overset{\vee}{\varphi}{}^{\theta}(B \cup C) - \overset{\vee}{\varphi}{}^{\theta}(B)\overset{\vee}{\varphi}{}^{\theta}(C)|$$

$$\leq \hat{P}(\hat{\xi}_t^B \cap \hat{\xi}_t^C \neq \emptyset \text{ for some } t)$$

$$\leq \sum_{x \in B} \sum_{x \in C} \hat{P}(\hat{N}_\infty^{\{x,y\}} = 1) .$$

Since $\bar{p}$ is transient, $\hat{P}(\hat{N}_\infty^{\{x,y\}} = 1) \to 0$ as $|y-x| \to \infty$. Hence, replacing C by $z + C$, we get (7.8), and the proof is finished. □

(7.9) Problem. Are there one-dimensional translation invariant voter models with equilibria other than $\delta_\emptyset$ and δ_{Z^d} ?

Holley and Liggett (1975) discuss voter models in more detail. Letting $\mathcal{I}_e$ denote the set of extreme invariant measures for a given irreducible model, they prove $\mathcal{I}_e = \{\delta_\emptyset, \delta_{Z^d}\}$ if $\bar{p}$ is recurrent, while $\mathcal{I}_e = \{\nu_\theta, 0 \leq \theta \leq 1\}$ when $\bar{p}$ is transient. The argument for the recurrent case is simple enough that we can give it here. Namely, if $\mu \in \mathcal{I}$ then using duality,

$$\varphi^\mu(x) = \varphi_t^\mu(x) = \sum_{y \in Z^d} p_t(x,y)\varphi^\mu(y) \qquad x \in Z^d, \quad t \in T,$$

where $p_t(x,y) = P(\xi_t^x = y)$. Thus $\varphi^\mu(x)$ is a harmonic function for the one particle random walk. By the Choquet-Deny theorem $\varphi^\mu(x)$ is a constant function. Thus (7.3) applies, so that μ is a mixture of the extreme measures $\delta_\emptyset$ and δ_{Z^d}. The proof that $\mathcal{I}_e = \{\nu_\theta, 0 \leq \theta \leq 1\}$ in the transient case is one of the tour de forces of the theory of particle systems; we refer the reader to Holley and Liggett (1975). In both the recurrent and transient cases, they also give necessary and sufficient conditions for an arbitrary $\mu \in \mathfrak{M}$ to belong to the domain of attraction of a given invariant measure. In particular, they show that if $\bar{p}$ is transient and μ is any translation invariant ergodic initial measure, then $\mu P^t \to \nu_\theta$, where $\theta = 1 - \varphi^\mu(0)$. In addition, they treat nontranslation invariant voter models, where the dual systems are coalescing Markov chains.

The qualitative difference between the recurrent and transient voter models leads to different sorts of questions for the two cases. When $\bar{p}$ is recurrent, one seeks to understand the implications of convergence to a mixture of $\delta_\emptyset$ and δ_{Z^d}. In the recurrent case <u>clustering</u> takes place, so that interest centers on cluster description. Given $A \in S$, say that x and y are in the same cluster if they can be connected by a path in Z^d whose vertices are either entirely in A or entirely in A^c. Thus, the clusters of configuration A are the connected components of A or A^c. One relevant quantity for systems which cluster is the asymptotic mean cluster size. Let $C(A)$, $A \in S$, be given by

$$C(A) = \lim_{n \to \infty} \frac{(2n)^d}{|\{\text{clusters of } A \text{ in } b_n(0)\}|}$$

provided the limit exists (undefined otherwise). For the one-dimensional basic voter model starting from μ_θ, the asymptotic growth of $C(\xi_t^{\mu_\theta})$ can be derived explicitly. First we need a general result which states that mixing is preserved by local additive systems at any time $t < \infty$. Note that this fact does not carry over to the limit as $t \to \infty$, as can be seen from voter models in the recurrent case.

(7.10) Lemma. Let $\mathcal{P}$ be a local percolation substructure which satisfies (1.5), $\{(\xi_t^A)\}$ the additive system induced by $\mathcal{P}$. If $\mu \in \mathfrak{M}$ is mixing, then μP^t is mixing for each $t \in T$.

Proof. Fix $B, C \in S_0$, $t \in T$ and a mixing measure μ. By duality we need to show

$$\lim_{|z| \to \infty} |\hat{E}[\varphi^\mu(\hat{\xi}_t^{B \cup (z+C)})] - \hat{E}[\varphi^\mu(\hat{\xi}_t^B)]\,\hat{E}[\varphi^\mu(\hat{\xi}_t^{z+C})]| = 0 .$$

To do this we use the construction from the proof of Theorem (2.6). Thus, let $\hat{\mathcal{P}}_1$ and $\hat{\mathcal{P}}_2$ be independent copies of $\mathcal{P}$. In addition, let $\hat{\mathcal{P}}_2^z = z + \hat{\mathcal{P}}_2$ be the translate of $\mathcal{P}_2$ by $z \in Z^d$. Define $(\hat{\xi}_t^B)$ in terms of $\hat{\mathcal{P}}_1$ and $\hat{\xi}_t^{z+C}$ in terms of $\hat{\mathcal{P}}_2^z$. Note that $\hat{\xi}_t^{z+C} = z + \hat{\xi}_t^C$. Introduce $\hat{\tau}_L^z = \min\{t : d(\hat{\xi}_t^B, \hat{\xi}_t^{z+C}) \le L\}$, with L as in (I.2.6). Now make a copy of $(\hat{\xi}_t^{B \cup (z+C)})$ by letting the flow which starts from B use $\hat{\mathcal{P}}_1$ while the flow starting from $z + C$ uses $\hat{\mathcal{P}}_2^z$ until τ_L and $\hat{\mathcal{P}}_1$ thereafter. With this representation $\hat{E}[\varphi^\mu(\hat{\xi}_t^B)\varphi^\mu(\hat{\xi}_t^{z+C})] = \hat{E}[\varphi^\mu(\hat{\xi}_t^B)]\,\hat{E}[\varphi^\mu(\hat{\xi}_t^{z+C})]$ and $\hat{\xi}_t^{B \cup (z+C)} = \hat{\xi}_t^B \cup \hat{\xi}_t^{z+C}$ if $\tau_L^z > t$. Thus the above absolute difference is majorized by

$$\hat{E}[\varphi^\mu(\hat{\xi}_t^B \cup (z+\hat{\xi}_t^C)) - \varphi^\mu(\hat{\xi}_t^B)\varphi^\mu(z+\hat{\xi}_t^C)] + \hat{P}(\tau_L^z < t) .$$

As $|z| \to \infty$, the first term tends to 0 since μ is mixing, and the second term tends to 0 because P does not have influence from ∞. □

(7.11) Theorem. Let $\{(\xi_t^A)\}$ be the basic voter model on Z. For $\theta \in (0,1)$,

$$\lim_{t\to\infty} \frac{C(\xi_t^{\mu_\theta})}{\sqrt{t}} = \frac{\sqrt{\pi}}{2\theta(1-\theta)} \quad \text{in P-probability .}$$

Proof. Say that A has an edge at $x+\frac{1}{2}$, $A \in S$, $x \in Z$, if $A(x) \neq A(x+1)$. Since $\mu_\theta P^t$ inherits mixing from μ_θ (by Lemma (7.10)), Birkhoff's theorem yields

$$\lim_{n\to\infty} \frac{|\{\text{edges of } \xi_t^{\mu_\theta} \text{ in } [-n,n]\}|}{2n} = P(\xi_t^{\mu_\theta} \text{ has an edge at } \tfrac{1}{2})$$

$\mu_\theta P^t$ - a.s. Because of the linear nature of Z, $|\{\text{clusters of } A \text{ in } [-n,n]\}|$ differs from $|\{\text{edges of } A \text{ in } [-n,n]\}|$ by at most 1. Since $\mu_\theta P^t$ has a positive density of edges for $\theta \in (0,1)$,

$$\lim_{n\to\infty} \frac{2n}{|\{\text{edges of } \xi_t^{\mu_\theta} \text{ in } [-n,n]\}|} = C(\xi_t^{\mu_\theta}) .$$

It follows that $C(\xi_t^{\mu_\theta}) = [P(\xi_t^{\mu_\theta}(0) \neq \xi_t^{\mu_\theta}(1))]^{-1}$ $\mu_\theta P^t$ - a.s. Computations similar to some from the proof of Theorem (7.2) yield

$$\begin{aligned} &P(\xi_t^{\mu_\theta}(0) \neq \xi_t^{\mu_\theta}(1)) \\ &= 2[P(0 \notin \xi_t^{\mu_\theta}) - P(\xi_t^{\mu_\theta} \cap \{0,1\} = \emptyset)] \\ &= 2[(1-\theta) - ((1-\theta) - \theta(1-\theta)\hat{P}(\hat{N}_t^{\{0,1\}} = 2))] \\ &= 2\theta(1-\theta)\hat{P}(\hat{N}_t^{\{0,1\}} = 2) . \end{aligned}$$

The last probability is the probability that a continuous time simple random walk with mean $-\frac{1}{2}$ holding times stays positive until time t given that it starts at 1. Using the reflection principle and the local central limit theorem, we have

$$\hat{P}(\hat{N}_t^{\{0,1\}} = 2) \sim \frac{1}{\sqrt{\pi t}} \qquad \text{as } t \to \infty .$$

The desired result follows. □

When $\bar{p}$ is transient, interest centers on the equilibria ν_θ, $\theta \in (0,1)$, for the voter model. Given a translation invariant measure $\mu \in \mathfrak{M}$, let ξ be μ-distributed, and set

$$S_n(\xi) = \sum_{x \in b_n(0)} (\xi(x) - E[\xi(x)]) .$$

μ is said to have <u>weak correlations</u> if

$$\lim_{n \to \infty} \frac{\mathrm{Var}(S_n(\xi))}{n^d} < \infty ,$$

and <u>strong correlations</u> if

$$\lim_{n \to \infty} \frac{\mathrm{Var}(S_n(\xi))}{n^d} = \infty .$$

(For reasonable measures μ the above limit exists.) Product measures, and more generally, measures with exponentially decaying correlations have weak correlations. For this class one can prove a central limit theorem of the usual sort: $n^{-d/2} S_n(\xi)$ converges in distribution to a mean - 0 normal random variable. Thus, for example, the unique equilibrium ν of any translation invariant extralineal additive system satisfies this central limit theorem, by virtue of Theorem (2.6). One of the most interesting properties of the ν_θ, $\theta \in (0,1)$, for a "transient" voter model is that they have strong correlations. We prove this for the basic voter model on Z^3.

(7.12) <u>Theorem</u>. Let $\{(\xi_t^A)\}$ be the basic 3-dimensional voter model. If ξ is ν_θ-distributed, $\theta \in (0,1)$, then

$$\lim_{n \to \infty} \frac{\mathrm{Var}(S_n(\xi))}{n^5} = \frac{3\gamma\theta(1-\theta)}{4\pi} \int\int_{u,v \in B_1(0)} \frac{du\,dv}{|u-v|} ,$$

where γ is the probability that simple random walk on Z^3 starting at 0 never returns to 0 ($\gamma \approx .65046267$), and $B_1(0)$ is the cube of side 2 in R^3 centered at 0.

<u>Proof</u>. The familiar formula for the variance of a sum yields

$$\mathrm{Var}(S_n(\xi)) = \sum\sum_{x,y \in b_n(0)} [E[\xi(x)\xi(y)] - E[\xi(x)]\,E[\xi(y)]] .$$

For $x \neq y$ we compute:

$$E[\xi(x)\xi(y)] - E[\xi(x)]\,E[\xi(y)]$$

$$= E[(1-\xi(x))(1-\xi(y))] - E[(1-\xi(x))]\,E[(1-\xi(y))]$$

$$= \hat{E}[\hat{N}_\infty^{\{x,y\}}] - (1-\theta)^2$$

$$= (1-\theta)\hat{P}(\hat{N}_\infty^{\{x,y\}} = 1) + (1-\theta)^2(1-\hat{P}(\hat{N}_\infty^{\{x,y\}} = 1)) - (1-\theta)^2$$

$$= \theta(1-\theta)\hat{P}(\hat{N}_\infty^{\{x,y\}} = 1)\,.$$

For $x = y$, $E[\xi(x)\xi(y)] - E[\xi(x)]\,E[\xi(y)] = \theta(1-\theta)$. Thus,

$$\mathrm{Var}(S_n(\xi)) = \theta(1-\theta)\,[(2n)^3 + \sum_{\substack{x,y \in b_n(0) \\ x \neq y}} \hat{P}(\hat{N}_\infty^{\{x,y\}} = 1)]\,.$$

Recall that $\hat{P}(\hat{N}_\infty^{\{x,y\}} = 1)$ is simply the probability that a random walk governed by $\bar{p}$ and starting from $x-y \in Z^3$ ever hits the origin. Since $\{(\xi_t^A)\}$ is the basic voter model, $\bar{p} = p$, and this probability is known to be asymptotically

(7.13) $$\hat{P}(\hat{N}_\infty^{\{x,y\}} = 1) \sim \frac{3\gamma}{4\pi}\,\frac{1}{|x-y|} \quad \text{as} \quad |x-y| \to \infty\,.$$

Therefore,

$$\frac{\mathrm{Var}(S_n(\xi))}{n^5} \approx \frac{3\gamma\,\theta(1-\theta)}{4\pi}\,n^{-5} \sum_{\substack{x,y \in b_n(0) \\ x \neq y}} \frac{1}{|x-y|}\,.$$

Make the change of variables $u = \frac{x}{n}$, $v = \frac{y}{n}$ to get

$$\frac{\mathrm{Var}(S_n(\xi))}{n^5} \approx \frac{3\gamma\,\theta(1-\theta)}{4\pi}\,n^{-6} \sum_{u,v \in B_1(0)} \frac{1}{|u-v|}\,.$$

The right side is a Riemann approximation to the desired integral. Further details are left to the reader. □

A more careful analysis of ν_θ leads to a central limit theorem in spite of the strong correlations: it can be proved that $n^{-5/2}\,S_n(\xi)$ converges in distribution to a normal variable. An even more interesting development deals with the "macroscopic dependency structure" of ξ. Define $S_n'(\xi)$ in terms of the block sum of side $2n$ centered at $2kn$ for some $k \neq 0$. Then S_n and S_n' have an asymptotic non-zero correlation as $n \to \infty$. This leads to a limiting field ξ_∞ if ξ is "renormalized" properly. See Bramson and Griffeath (1978a) for more on renormalizing the voter model.

(7.14) Notes. Voter models were studied independently by Clifford and Sudbury (1973) and Holley and Liggett (1975). Clifford and Sudbury discovered the qualitative dichotomy between the recurrent and transient cases, whereas Holley and Liggett first determined the structure of $\mathcal{J}$. Theorem (7.11) is from Bramson and Griffeath (1978b); nothing is known about the rate of clustering when $d = 2$. R. Arratia supplied the proof of Lemma (7.10) (private communication). A proof of the central limit theorem for random fields with exponentially decaying correlations may be found in Malyshev (1975). Theorem (7.12) is due to Bramson and Griffeath (1978a). See Dawson (1978), Dawson and Ivanoff (1978), Durrett (1978) and Fleischmann (1978) for related results in other contexts. The asymptotics for random walks used in this section (e.g. (7.13)) are derived by Spitzer (1976).

8. Biased voter models.

The voter models of the previous section were symmetric in 0's and 1's, since they satisfied $c_x(A) = c_x(A^c)$. Similar systems with a uniform asymmetry are called "biased voter models." Without loss of generality we assume that the bias favors 1's, and define the biased voter model on Z^d with parameter $\lambda > 1$ and probability density p to be the spin system $\{(\xi_t^A)\}$ with flip rates

$$c_x(A) = A(x) + (\lambda - (1+\lambda)A(x)) \sum_{z \in A} p_{z-x} .$$

In this model the voter at x is influenced by voter y with weight λp_{y-x} if x is "against" and y is "for", but only with weight p_{y-x} if x is "for" and y is "against." For simplicity, throughout the rest of this section we restrict attention to the basic cases, where

$$p_z = \frac{1}{2d} \quad |z| = 1 \qquad\qquad (p_z = 0 \quad \text{otherwise.})$$

The percolation substructure for $\{(\xi_t^A)\}$ may be described schematically as:

$$\delta \overset{x}{\longleftarrow}\!\!\overset{x+z}{\rule{0pt}{0pt}} \quad \text{at rate } \frac{1}{2d}$$

$$(|z| = 1) \ .$$

$$\overset{x}{\longleftarrow}\!\!\overset{x+z}{\rule{0pt}{0pt}} \quad \text{at rate } \frac{\lambda-1}{2d}$$

Thus the dual system $\{(\hat{\xi}_t^A)\}$ is a coalescing branching system with a random walk part and a nearest neighbor branching part: $\hat{P}$ has

$$\delta \overset{x}{\longrightarrow}\!\!\overset{x+z}{\rule{0pt}{0pt}} \quad \text{at rate } \frac{1}{2d}$$

$$(|z| = 1) \ .$$

$$\overset{x}{\longrightarrow}\!\!\overset{x+z}{\rule{0pt}{0pt}} \quad \text{at rate } \frac{\lambda-1}{2d}$$

We now show how $\{(\hat{\xi}_t^A)\}$ may be used to gain information about the equilibria for $\{(\xi_t^A)\}$.

(8.1) Theorem. The only extreme invariant measures for the basic d-dimensional biased voter model are $\delta_\emptyset$ and δ_{Z^d} .

Proof. We need to show that if $\mu \in \mathcal{I}$, then $\mu = c\delta_\emptyset + (1-c)\delta_{Z^d}$, or equivalently that $\varphi^\mu(\Lambda) = c$ whenever $\Lambda \neq \emptyset$, for some constant c . To do so, observe that $M_t = \varphi^\mu(\hat{\xi}_t^\Lambda)$ is a martingale for given $\mu \in \mathcal{I}$, $\Lambda \in S_0$, since

$$E[M_t] = \varphi^{\mu P^t}(\Lambda) = \varphi^\mu(\Lambda) = M_0 \ .$$

Clearly M_t is bounded, so if $\hat{\tau}(\Lambda,\Lambda') = \min\{t : \hat{\xi}_t^\Lambda \supset \Lambda'\}$ is finite with $\hat{P}$-probability one for $\Lambda \neq \emptyset$, then by the martingale stopping theorem and monotonicity of $\{(\xi_t^A)\}$,

$$\varphi^\mu(\Lambda) = M_0 = E[\varphi^\mu(\hat{\xi}^\Lambda_{\hat{\tau}(\Lambda,\Lambda')})] \le \varphi^\mu(\Lambda') \ .$$

Reversing the roles of Λ and Λ' we see that φ^μ is constant on $S_0 - \{\emptyset\}$. It therefore remains only to check that

$$\hat{P}(\hat{\xi}_t^\Lambda \supset \Lambda' \text{ for some } t) = 1 \qquad \forall \Lambda,\Lambda' \in S_0 - \{\emptyset\} \ .$$

Now it is quite clear that

$$\hat{P}(\hat{\xi}_t^0 \supset \Lambda' \text{ for some } t) > 0 \qquad \forall \Lambda' \in S_0 ,$$

since the random walk and branching mechanisms make a box covering Λ' accessible from $\{0\}$. By monotonicity and a standard Markov process argument, it therefore suffices to show

$$P(0 \in \xi_t^{\Lambda} \text{ for arbitrarily large } t) = 1 \qquad \forall \Lambda \in S_0 - \{\emptyset\} .$$

This last property is verified by finding a recurrent Markov chain (X_t) on Z^d such that $X_t \subset \xi_t^{\Lambda}$ for all t. To define (X_t), we start at some $z \in \Lambda$ and follow a random walk arrow whenever it occurs, but a branching arrow only if it takes us closer to 0. The resulting chain is obviously imbedded in (ξ_t^{Λ}), since it follows a path up in $\mathcal{P}$. It is recurrent since it has a drift toward 0 caused by the selective following of the branching. A rigorous proof of recurrence can be based on the fact that $(|X_{t \wedge \tau_b}|^2)$ is a supermartingale, where τ_b is the hitting time of a sufficiently large d-sphere centered at 0. Details are left to the reader. □

It is intuitively clear that if a biased voter process (ξ_t^A) does not die out, then its configuration will converge to Z^d as $t \to \infty$. In one dimension this is quite easy to prove.

(8.2) Proposition. Let $\{(\xi_t^A)\}$ be the basic biased voter model on Z. Then for any $\mu \in \mathfrak{M}$,

$$\lim_{t \to \infty} \xi_t^{\mu} = Z \qquad P-\text{a.s. on } \{\tau_{\emptyset}^{\mu} = \infty\} .$$

Proof. Note that ξ_t^x is a block $[L_t, R_t]$, where L_t and R_t evolve like independent random walks with drift $\lambda - 1 > 0$ toward $-\infty$ and $+\infty$ respectively, whenever $R_t - L_t > 0$. The claim follows easily for $\mu = \delta_x$. By additivity the result holds for any measure δ_A, $A \in S_0$. A simple approximation argument extends the result to any δ_A, $A \in S$, and hence to arbitrary $\mu \in \mathfrak{M}$. □

As an immediate consequence we obtain the weaker convergence result:

(8.3) Corollary. For the basic biased voter model on Z,

$$\mu P^t \to P(\tau_{\emptyset}^{\mu} < \infty)\, \delta_{\emptyset} + P(\tau_{\emptyset}^{\mu} = \infty)\, \delta_Z \qquad \forall \mu \in \mathfrak{M} .$$

The law of large numbers shows that ξ_t^x grows linearly given nonextinction when $d = 1$. For $d \geq 2$, the analogue of Proposition (8.2) still holds, and in fact $(\xi_t^0 \mid \{\tau_\emptyset^0 = \infty\})$ becomes essentially a solid "blob" whose radius grows linearly with t. In the limiting case of total bias ("$\lambda = \infty$"), Richardson (1973) has proved such a theorem. He shows that there is a norm $\| \ \|$ on R^d such that

$$\forall \varepsilon > 0 \quad \exists t_0 < \infty : P(B_{(1-\varepsilon)t} \subset \xi_t^0 \subset B_{(1+\varepsilon)t}) \geq 1 - \varepsilon \qquad \forall t > t_0 ,$$

where $B_r = \{x \in Z^d : \|x\| \leq r\}$. Analogous results for the models with $1 < \lambda < \infty$ are the subject of two forthcoming papers by Bramson and Griffeath.

(8.4) Notes. The basic biased voter models restricted to S_0 were introduced by Williams and Bjerknes (1972) as models for cancer growth. Their simulations and conjectures led to a great deal of work on growth rates of S_0-valued particle systems; see Mollison (1977) for a survey of these and related problems. Schwartz (1977) has studied the basic biased voter models on S, and is the source of Theorem (8.1).

9. Coalescing random walks.

The lineal additive systems known as coalescing random walks have already appeared as duals for voter models. Such a system is determined by a probability density $p = (p_y ; y \in Z^d)$; its percolation substructure has the representation:

$$I = Z^d \times Z^d , \quad \lambda_{z,x} = p_{z-x} , \quad V_{z,x} \equiv \emptyset ;$$

$$W_{z,x}(y) = \{z\} \text{ if } y = x \quad (= \{y\} \text{ otherwise}).$$

The intuitive description was given in Example (I.1.1) and in Section (II.7). In this section we reverse perspectives, and derive some properties of coalescing random walks by using the voter model as an auxilliary system.

To begin, there is the question of ergodicity. One expects a limit measure $\delta_\emptyset$ starting from any initial state, since extant particles become more and more rare. This is confirmed by our first result.

(9.1) <u>Proposition</u>. Let $\{(\xi_t^A)\}$ be the coalescing random walks with transition density p. Then

$$\mu P^t \to \delta_\emptyset \quad \text{as} \quad t \to \infty \qquad \forall \mu \in \mathfrak{M} .$$

<u>Proof</u>. By monotonicity, it suffices to check that

$$\lim_{t \to \infty} P(\xi_t^{Z^d} \ni x) = 0 \qquad \forall x \in Z^d .$$

The dual system is the corresponding voter model, and

$$P(\xi_t^{Z^d} \ni x) = \hat{P}(\hat{\xi}_t^x \neq \emptyset) .$$

If $\hat{\xi}_t^x = \Lambda$, $|\Lambda| = k$, the cardinality of the voter process becomes $k+1$ at rate $q(\Lambda) = \sum_{y \in \Lambda^c} \sum_{z \in \Lambda} p_{z-y} \geq 1$, and becomes $k-1$ at rate $\sum_{y \in \Lambda} \sum_{z \in \Lambda^c} p_{z-y} = q(\Lambda)$. Thus $(|\hat{\xi}_t^x|)$ jumps like at simple random walk with absorption at 0 after exponential holding times with rate at least 2. Hence $\hat{P}(\hat{\tau}_\emptyset^x < \infty) = 1$, and the proposition follows. □

Since the distribution of $\xi_t^{Z^d}$ converges to $\delta_\emptyset$, it is natural to ask whether an individual site, the origin say, is visited by the coalescing random walks at arbitrarily large times, or whether there is a last visit. This "recurrence" question is settled by our next result.

(9.2) <u>Theorem</u>. If $\{(\xi_t^A)\}$ is the coalescing random walks on Z^d with density p, then

$$P(\limsup_{t \to \infty} \xi_t^{Z^d}(0) = 1) = 1 .$$

The proof relies on a lemma, which will imply that the expected amount of time that 0 is occupied by $(\xi_t^{Z^d})$ is infinite. This preliminary result, stated for the voter model, is of interest in its own right.

(9.3) <u>Lemma</u>. Let $\{(\hat{\xi}_t^A)\}$ be the voter model on Z^d with density p. Then for $t \geq 0$,

$$\hat{P}(\hat{\xi}_t^0 \neq \emptyset) \geq (1+t)^{-1} .$$

Proof. As already noted, when $\hat{\xi}_t^0 = \Lambda$, $1 \leq |\Lambda| < \infty$, the voter process increases or decreases by one particle at the same exponential rate $q(\Lambda) = \sum'_{y \in \Lambda^c} \sum_{z \in \Lambda} p_{z-y}$. Clearly $q(\Lambda) \leq |\Lambda|$. Let $(Z_t)_{t \geq 0}$ be a birth and death process on $\{0,1,2,\cdots\}$ with absorption at 0, and with transition from k to $k-1$ or $k+1$ $(k \geq 1)$ at the same exponential rate k. Start (Z_t) at 1, and note that since this process jumps at least as fast as $(\hat{\xi}_t^0)$, it will be absorbed more quickly. Thus

$$P(\hat{\xi}_t^0 \neq \emptyset) \geq P(Z_t \neq 0) .$$

Let $u(t)$ denote the right side of this last inequality. Then $u(0) = 1$ and $\frac{du}{dt} = -u^2$ (Problem (9.4).) Hence $u(t) = (1+t)^{-1}$, completing the proof. □

(9.4) Problem. Show that the function $u(t)$ defined above satisfies $\frac{du}{dt} = -u^2$. (Hint: (Z_t) is a Galton-Watson process.)

Proof of Theorem (9.2). Let $\tau_t = \min\{s \geq t : 0 \in \xi_s^{Z^d}\}$, and note that $\{\limsup_{t\to\infty} \xi_t^{Z^d}(0) = 1\} = \lim_{t\to\infty} \lim_{u\to\infty} \{\tau_t \in [t,u]\}$. For $0 \leq t < u$, by the Markov property and monotonicity,

$$E[\int_t^u \xi_s^{Z^d}(0)\,ds] = \int_t^u \int_S P(\tau_t \in dr, \xi_{\tau_t}^{Z^d} \in dA) E[\int_0^{u-r} \xi_s^A(0)\,ds]$$

$$\leq \int_t^u \int_S P(\tau_t \in dr, \xi_{\tau_t}^{Z^d} \in dA) E[\int_0^u \xi_s^{Z^d}(0)\,ds]$$

$$= P(\tau_t \in [t,u]) E[\int_0^u \xi_s^{Z^d}(0)\,ds] .$$

Thus

$$P(\tau_t \in [t,u]) \geq \frac{E[\int_t^u \xi_s^{Z^d}(0)\,ds]}{E[\int_0^u \xi_s^{Z^d}(0)\,ds]} .$$

For each fixed t, the left side tends to 1 as $u \to \infty$, since by duality and the last lemma,

$$E[\int_0^\infty \xi_s^{Z^d}(0)\,ds] = \int_0^\infty P(0 \in \xi_s^{Z^d})\,ds$$

$$= \int_0^\infty P(\hat{\xi}_s^0 \neq \emptyset)\,ds \geq \int_0^\infty (1+s)^{-1}\,ds = \infty \quad .$$

Thus $P(\limsup_{t\to\infty} \xi_t^{Z^d}(0) = 1) = \lim_{t\to\infty} \lim_{u\to\infty} P(\tau_t \in [t,u]) = 1$, and the proof is finished. $\square$

With more care one can extend Theorem (9.2) to coalescing random walks starting from any dense configuration A. The real content of the theorem lies in the case where p is transient; there is much simpler and more general result if p is recurrent:

(9.5) Problem. Show that for p irreducible recurrent,

$$P(\limsup_{t\to\infty} \xi_t^A(0) = 1) = 1 \qquad \forall A \neq \emptyset \; .$$

(p irreducible means that the group generated by $\{y \in Z^d : p_y > 0\}$ is all of Z^d .)

We conclude this section with cluster size results for coalescing random walks on Z. To keep matters simple, we start from δ_Z. Let $D_+(\xi_t^Z)$ be the distance from the origin to the first non-negative site x such that $x \in \xi_t^Z$. We prove a distribution limit theorem for $D_+(\xi_t^Z)$.

(9.6) Theorem. With D_+ defined as above,

$$\lim_{t\to\infty} P\left(\frac{D_+(\xi_t^Z)}{\sqrt{t}} \leq \alpha\right) = \frac{1}{\sqrt{\pi}} \int_0^\alpha e^{-\frac{s^2}{4}}\,ds \; .$$

Proof. Write $n = n(t) = \lfloor \alpha\sqrt{t} \rfloor$. ($\lfloor s \rfloor$ is the greatest integer in s .) By duality,

$$P\left(\frac{D_+(\xi_t^Z)}{\sqrt{t}} \leq \alpha\right) = P(\xi_t^Z \cap [0,n] \neq \emptyset)$$

$$= \hat{P}(\hat{\xi}_t^{[0,n]} \neq \emptyset) \; .$$

The key observation is that $(|\hat{\xi}_t^{[0,n]}| - 1)$ is a simple random walk on $\{0,1,\cdots\}$ with jump rate 2, starting at n, and with absorption at -1. Call this process (Y_t). The reflection principle for random walk states that if (X_t) is simple random walk on Z with jump rate 2, starting at 0, then

$$P(Y_t \geq 0) = P(X_t \leq n) - P(X_t \geq n+2)$$
$$= P\left(\frac{X_t}{\sqrt{2t}} \leq \frac{\lfloor \alpha\sqrt{t} \rfloor}{\sqrt{2t}}\right) - P\left(\frac{X_t}{\sqrt{2t}} \geq \frac{\lfloor \alpha\sqrt{t} \rfloor + 2}{\sqrt{2t}}\right) .$$

According to the central limit theorem, the right side converges to

$$\frac{1}{\sqrt{2\pi}} \left(\int_{-\infty}^{\frac{\alpha}{\sqrt{2}}} - \int_{\frac{\alpha}{\sqrt{2}}}^{\infty} \right) e^{-\frac{u^2}{2}} du = \sqrt{\frac{2}{\pi}} \int_0^{\frac{\alpha}{\sqrt{2}}} e^{-\frac{u^2}{2}} du .$$

Since $P\left(\frac{D_+(\xi_t^Z)}{\sqrt{t}} \leq \alpha\right) = P(Y_t \geq 0)$, the change of variables $s = \sqrt{2}\, u$ gives the desired result. □

(9.7) Problems. Define the mean interparticle distance $D(\xi_t^Z)$ as a limit of the mean interparticle distances of $\xi_t^Z|_{b_n(0)}$. Use the method of Theorem (7.11) to prove that

(9.8) $$\lim_{t\to\infty} \frac{D(\xi_t^Z)}{\sqrt{t}} = \sqrt{\pi} \quad \text{in } P\text{-probability.}$$

Show that $E\left[\frac{D_+(\xi_t^Z)}{\sqrt{t}}\right] \to \frac{2}{\sqrt{\pi}}$ as $t\to\infty$, so that if $D_0(\xi_t^Z)$ is the distance between the particles immediately surrounding the origin in ξ_t^Z, then

(9.9) $$\lim_{t\to\infty} E\left[\frac{D_0(\xi_t^Z)}{\sqrt{t}}\right] = \frac{4}{\sqrt{\pi}} .$$

Explain the discrepancy between the constants in (9.8) and (9.9). Show that (9.8) continues to hold if (ξ_t^Z) is replaced by (ξ_t^μ), where μ is any translation invariant mixing measure except $\delta_\emptyset$. Show that (9.9) continues to hold if (ξ_t^Z) is replaced by (ξ_t^μ), where μ is any translation invariant measure with $\int D_0(A)\mu(dA) < \infty$.

(9.10) Notes. Coalescing random walks were introduced by Holley and Liggett (1975) as dual processes for voter models. They are studied in their own right by Griffeath (1978b); Proposition (9.1) and Theorem (9.2) are from that paper. Theorem (9.6) and the results in Problems (9.7) are due to Bramson and Griffeath (1978b), who also obtain similar theorems for the basic voter model on Z .

10. Stirring and exclusion systems.

Our final section on additive systems deals with a class of models called random stirrings. For simplicity we discuss only the translation invariant situation. Let $W_{i,0}$, $i \in I_0$, be permutations of Z^d which leave all but a finite number of sites fixed. The rate for $W_{i,0}$ is $\lambda_{i,0}$, where $\sum_i \lambda_{i,0} < \infty$. Let $\mathcal{P}$ be the translation invariant lineal percolation substructure determined by the $\lambda_{i,0}$ and $W_{i,0}$. Then the additive system $\{(\xi_t^A)\}$ induced by $\mathcal{P}$ is a random stirring. Clearly the dual $\{(\hat{\xi}_t^B)\}$ for any such system is another random stirring, namely the one constructed from permutations $\hat{W}_{i,x} = W_{i,x}^{-1}$ = the inverse of $W_{i,x}$. It is also easy to see that the product measures μ_θ are equilibria. In fact, since $|\hat{\xi}_t^B| = |B|$ for all $B \in S_0$, $t \geq 0$, by (1.11),

$$\varphi_t^{\mu_\theta}(B) = (1-\theta)^{|B|} = \varphi^{\mu_\theta}(B) .$$

We now derive a convergence theorem for random stirrings. It states that any measure μ with density θ which satisfies a certain mixing condition is in the domain of attraction of μ_θ.

(10.1) Theorem. Let $\{(\xi_t^A)\}$ be a random stirring. Given $\mu \in \mathfrak{M}$, assume that $\varphi^\mu(x) \equiv 1-\theta$, and that

$$\lim_{R\to\infty} \sup_{\Lambda \in S_0^{n,R}} \left| \varphi^\mu(\Lambda) - \prod_{x\in\Lambda} \varphi^\mu(x)\right| = 0 , \tag{10.2}$$

where $S_0^{n,R} = \{\Lambda \in S_0 : |\Lambda| = n \text{ and } y-x \geq R, \ \forall x, y \in \Lambda, y \neq x\}$. Then $\mu P^t \to \mu_\theta$ as $t \to \infty$.

Proof. By duality equation (1.10), for each $B \in S_0$,

$$|\varphi_t^\mu(B) - \varphi^{\mu_\theta}(B)| = |\hat{E}[\varphi^\mu(\hat{\xi}_t^B) - (1-\theta)^{|B|}]|$$

$$\leq \sup_{\Lambda \in S_0^{|B|,R}} |\varphi^\mu(\Lambda) - (1-\theta)^{|\Lambda|}| + P(\hat{\xi}_t^B \notin S_0^{|B|,R}) .$$

The hypotheses on μ state that the above supremum tends to 0 as $R \to \infty$. To prove the theorem, it therefore suffices to check that

$$(10.3) \qquad \lim_{t\to\infty} P(\hat{\xi}_t^B \not\subset S_0^{|B|,R}) = 0 \quad \text{for each } R .$$

Rewrite this last probability as

$$P(\exists x,y \in B,\ x \neq y : |\hat{\xi}_t^y - \hat{\xi}_t^x| < R)$$

$$\leq \sum_{\substack{x,y \in B \\ x \neq y}} P(|\hat{\xi}_t^y - \hat{\xi}_t^x| < R),$$

so that (10.3) will follow from the fact that

$$(10.4) \qquad \lim_{t\to\infty} P(|\hat{\xi}_t^y - \hat{\xi}_t^x| < R) = 0 \qquad \forall x,y \in Z^d .$$

To get (10.4), define Markov chains $\{(X_t^x)\}$ on Z^d by $X_t^{y-x} = \hat{\xi}_t^y - \hat{\xi}_t^x$. Counting measure is invariant for $\{(X_t^x)\}$, so the chains are null recurrent or transient. In either case $\lim_{t\to\infty} P(|X_t^{y-x}| < R) = 0$, so the proof is finished. □

The most important random stirrings are the <u>additive exclusion systems</u>, where each $W_{i,x}$ permutes exactly two sites. These models may be interpreted as systems of particles performing independent random walks, but subject to an <u>exclusion rule</u>: whenever a particle attempts to jump to a site which is already occupied it is not allowed to do so. (Note that (ξ_t^x) is <u>not</u> the motion of an individual particle under this interpretation.) Liggett has studied the additive exclusion systems, and a number of more general models, in a long series of beautiful papers. The reader is referred to Liggett (1977) for a self-contained survey of his work. One basic theorem states that the extreme equilibria for any additive exclusion system are precisely $\{\mu_\theta ;\ 0 \leq \theta \leq 1\}$. Also, a result similar to (10.1) states that $\mu P^t \to \mu_\theta$ as $t \to \infty$ whenever μ is translation invariant and (spatially) ergodic with density θ.

(10.5) <u>Problem</u>. Show that any additive exclusion system is self-dual. Which other random stirrings are self-dual?

(10.6) Notes. Lee (1974) introduced random stirrings and proved Theorem (10.1) for systems on Z . He also considered analogous processes on R . Harris (1976) has additional remarks on stirrings. Spitzer (1970) first formulated exclusion models, and proved self-duality in the additive case. A detailed analysis was subsequently carried out by Liggett (1973, 1974, 1975, 1976) and by Spitzer (1974a).

CHAPTER III: CANCELLATIVE SYSTEMS

1. The general construction.

This chapter is devoted to a second class of particle systems which, like the additive ones, can be defined by means of percolation substructures. Given $\mathcal{P} = \mathcal{P}(\lambda; V, W)$, if we define

(1.1) $$\eta_t^A = \{x : N_t^A(x) \text{ is odd}\} ,$$

then $\{(\eta_t^A)\}$ is another S-valued Markov family, called the (canonical) cancellative particle system induced by $\mathcal{P}$. If $\eta_t^A = B$ and the (i,x) clock goes off, then (1.1) implies that configuration B jumps to

$$\mathfrak{C}_{i,x}(B) = [\underset{y \in B}{\Delta} W_{i,x}(y)] \Delta V_{i,x} .$$

(Since $W_{i,x}(y) = \{y\}$ for all but a finite number of sites y, this symmetric difference makes sense even when $B \in S_\infty$.) If $\mathcal{P}$ has no influence from ∞, then a result similar to Proposition (II.1.4) ensures that $\{(\eta_t^A)\}$ is Feller. The conditions (II.1.5) and (II.1.6) are again sufficient, and will apply to all of the specific models in this chapter. Our first result for cancellative systems is the obvious analogue of (II.1.2).

(1.2) Proposition. If $\{(\eta_t^A)\}$ is a cancellative system, then

$$\eta_t^{A \Delta B} = \eta_t^A \Delta \eta_t^B \Delta \eta_t^\emptyset \qquad A, B \in S, \quad t \geq 0 .$$

Proof. $x \in \eta_t^A \Delta \eta_t^B \Delta \eta_t^\emptyset \iff N_t^A(x) + N_t^B(x) + N_t^\emptyset(x) \equiv 1 \pmod 2$

$\iff (N_t^A(x) - N_t^{A \cap B}(x)) + (N_t^B(x) - N_t^{A \cap B}(x)) + N_t^\emptyset(x) \equiv 1 \pmod 2$

$\iff N_t^{A \Delta B}(x) \equiv 1 \pmod 2 \iff x \in \eta_t^{A \Delta B}$. □

Unlike the additive systems, the models we will now study are typically not monotone. This is clearly the case for one of the simplest cancellative systems: the annihilating random walks of Example (I.1.3). Another example is given in the following exercise.

(1.3) Problem. Let p be a transition density on Z^d. The anti-voter model determined by p and $\lambda > 0$ is the spin system $\{(\eta_t^A)\}$ with flip rates

$$c_x(A) = \lambda \, [\, A(x) + (1-2A(x)) \sum_{z \in A^c} p_{z-x} \,] \ .$$

Show that such a model is cancellative, and determine $\mathcal{P}$.

Due to the lack of monotonicity, new techniques will have to be developed. Fortunately there is still a duality equation, although the general dual system $\{(\hat{\eta}_t^A)\}$ is more complicated than in the additive setting. We now proceed to develop this duality theory. General ergodicity results will be proved in the next section, and then several specific cancellative systems will be studied in some detail.

Let $\{(\eta_t^A)\}$ be defined by (1.1) for given $\mathcal{P}(\lambda; V, W)$. The dual processes $(\tilde{\eta}_t^B)$, $B \in S_0$, will have state space $\tilde{S} = (S_0 \times \{0,1\}) \cup \{\Delta\}$, Δ an isolated point. The second factor space $\{0,1\}$ is necessary because parity plays an important role in cancellative systems. Say that the (i,x)-flow in $\mathcal{P}$ has pure births if

$$\lambda_{i,x} > 0 \,, \quad V_{i,x} \neq \emptyset \quad \text{and} \quad W_{i,x}(y) = \{y\} \ \forall y \ ;$$

for such (i,x), the labels β are called pure births. To define $(\tilde{\eta}_t^B)$, introduce the modification $\tilde{\mathcal{P}}(\tilde{\lambda}, V, \hat{W})$ of the dual substructure $\hat{\mathcal{P}}(\hat{\lambda}, V, \hat{W})$ such that

$$\begin{aligned} \tilde{\lambda}_{i,x} &= 2\lambda_{i,x} \quad \text{if} \quad (i,x) \text{ has pure births in } \hat{\mathcal{P}} \,, \\ &= \lambda_{i,x} \quad \text{otherwise} \,. \end{aligned}$$

Define

$$\tilde{\mathcal{B}}_t = \{(x,s),\ 0 < s \le t :\ \text{a pure birth occurs at } (x,s) \text{ in } \tilde{\mathcal{P}}\} \,,$$

$$\tilde{\tau}_\Delta^B = \inf\{t \ge 0 :\ \exists \text{ odd number of paths up from } (B,0) \text{ to } \tilde{\mathcal{B}}_t \text{ in } \tilde{\mathcal{P}}\}$$

Now let

$$(1.4) \qquad \begin{aligned} \tilde{\eta}_t^B &= (\hat{\eta}_t^B, \varepsilon_t^B) & 0 \le t < \tilde{\tau}_\Delta^B \\ &= \Delta & \tilde{\tau}_\Delta^B \le t < \infty \,, \end{aligned}$$

where

$$\hat{\eta}_t^B = \{x : \exists \text{ odd number of paths up from } (B,0) \text{ to } (x,t) \text{ in } \tilde{\mathcal{P}}\},$$

$$\varepsilon_t^B \equiv \text{number of paths up from } (B,0) \text{ to } \mathcal{B}_t \text{ in } \tilde{\mathcal{P}} \pmod 2 .$$

(Recall that $\mathcal{B}_t = \{(x,s),\ 0 < s \le t :$ a birth occurs at (x,s) in $\tilde{\mathcal{P}}\}$.) The duality equation for cancellative systems may be stated as follows.

(1.5) Theorem. Let $\{(\eta_t^B)\}$ be the cancellative system induced by $\mathcal{P}$, $\{(\tilde{\eta}_t^B);\ B \in S_0\}$ the collection of dual processes defined by (1.4) . Then for each $t \ge 0$, $A \in S$, $B \in S_0$,

$$P(|\eta_t^A \cap B| \text{ even}) = \tilde{P}(|\hat{\eta}_t^B \cap A| + \varepsilon_t^B \text{ even},\ \tilde{\tau}_\Delta^B > t) + \tfrac{1}{2}\tilde{P}(\tilde{\tau}_\Delta^B \le t) .$$

Proof. Let $\bar{\mathcal{P}}_t(\tilde{\lambda}\,;V,W)$ and $\tilde{\mathcal{P}}_t(\tilde{\lambda}\,;V,\hat{W})$ be the forward and reverse percolation substructures on $Z^d \times [0,t] = Z^d \times [\hat{0},\hat{t}]$ constructed as in Chapter I. A copy of $\mathcal{P}_t$, the restriction of $\mathcal{P}(\lambda\,;V,W)$ to $Z^d \times [0,t]$, can be embedded in $\bar{\mathcal{P}}_t$ as follows: at each location (x,s) where a pure birth occurs in $\bar{\mathcal{P}}_t$, flip a fair coin to decide whether to subscript the label β with a 0 or with a 1 . If all of the coin flips are independent, and only the β_1's are pure births in $\mathcal{P}_t$ (i.e. the β_0's are ignored), then we obtain the desired version of $\mathcal{P}_t$. Thus $(\eta_s^A)_{0 \le s \le t}$ and $(\tilde{\eta}_s^B)_{0 \le s \le t}$ are realized on the joint substructure. Moreover,

$$\{|\eta_t^A \cap B| \text{even},\ \tilde{\tau}_\Delta^B > t\} = \{|\hat{\eta}_t^B \cap A| + \varepsilon_t^B \text{ even},\ \tilde{\tau}_\Delta^B > t\}\ P\text{-a.s.},$$

since both events are a.s. equivalent to

$$\{\text{an even number of paths between } ((A,0) \cup \mathcal{B}_t) \text{ and } (B,t),\ \hat{\tau}_\Delta^B > t\} .$$

To see this, use the fact that

$$\begin{aligned} |\eta_t^A \cap B| \text{ even} &\iff |\{y \in B : N_t^A(y) \text{ odd}\}| \text{ even} \\ &\iff N_t^A(B) \text{ even} \\ &\iff |\hat{\eta}_t^B \cap A| + \varepsilon_t^B \text{ even}, \quad P\text{-a.s.} \end{aligned}$$

If $P(\tilde{\tau}_\Delta^B \le t) = 0$ we're done. Otherwise, write

$$\{|\eta_t^A \cap B| \text{ even}, \tilde{\tau}_\Delta^B \le t\} = (E \cap F \cap G) \cup (E \cap F^c \cap G^c),$$

where

$$E = \{\tilde{\tau}_\Delta^B \le t\}, \quad F = \{\beta_1\text{'s occur at } \tilde{\tau}_\Delta^B\},$$

$$G = \{\text{odd number of paths to } (B,t) \text{ from } ((A,0) \cup \mathcal{B}_t) - (Z^d, \tilde{\tau}_\Delta^B)\}.$$

The key observations, which follow from the construction, are that F and G are conditionally independent given E, and that $P(F|E) = \frac{1}{2}$. We conclude that

$$P(E \cap F \cap G) + P(E \cap F^c \cap G^c) = \frac{1}{2}[P(E \cap G) + P(E \cap G^c)]$$
$$= \frac{1}{2} P(E).$$

The desired duality equation follows immediately. □

The quantities $\psi_t^\mu(B) = P(|\eta_t^\mu \cap B| \text{ even})$ play the role for cancellative systems that the zero functions $\varphi_t^\mu(B)$ did in the additive case. In order to prove ergodic theorems using Theorem (1.5), we need to know that any $\mu \in \mathfrak{M}$ is uniquely determined by its cylinder set probabilities of the form $\psi^\mu(B) = \mu(\{A : |A \cap B| \text{ even}\})$. To see this, we observe that φ^μ is determined by ψ^μ according to the formula

$$\varphi^\mu(B) = 2^{-|B|} \sum_{\Lambda \subset B} [2\psi^\mu(\Lambda) - 1], \qquad B \in S_0. \tag{1.6}$$

(1.7) Problem. Derive equation (1.6).

The states $(\emptyset, 0)$ and $(\emptyset, 1)$ are both traps for $(\tilde{\eta}_t^B)$, since if

$$\text{number of paths up from } (B,0) \text{ to } (x,t) \text{ in } \tilde{\mathcal{P}} \text{ is even} \quad \forall x \in Z^d,$$

then for any $s \ge t$,

$$\text{number of paths up from } (B,0) \text{ to } (x,s) \text{ in } \tilde{\mathcal{P}} \text{ is even} \quad \forall x \in Z^d$$

P - a.s. State Δ is also a trap. We let $\tilde{\tau}_0^B$, $\tilde{\tau}_1^B$ and $\tilde{\tau}_\Delta^B$ be the respective hitting times; clearly at most one is finite. For convenience, we also put

$\tilde{\tau}^B = \tilde{\tau}_0^B \wedge \tilde{\tau}_1^B \wedge \tilde{\tau}_\Delta^B$. Two easy consequences of Theorem (1.5) conclude this section.

(1.8) Corollary. Let $\{(\eta_t^A)\}$ be cancellative. There is an invariant measure $\nu \in \mathfrak{M}$ such that

$$\mu_{\frac{1}{2}} P^t \to \nu \qquad \text{as} \qquad t \to \infty ,$$

where $\psi^\nu(B) = \frac{1}{2}[\tilde{P}(\tilde{\tau}_0^B < \infty) + \tilde{P}(\tilde{\tau}_1^B = \infty)]$.

Proof. Integrating the duality equation with respect to $\mu_{\frac{1}{2}}$, and using the fact $\psi^{\mu_{\frac{1}{2}}}(B) = \frac{1}{2}$ whenever $\emptyset \neq B \in S_0$, we get

$$\psi_t^{\mu_{\frac{1}{2}}}(B) = \tilde{E}[\mu_{\frac{1}{2}}(\{A : |A \cap \hat{\eta}_t^B| + \varepsilon_t^B \text{ even}\}), \tilde{\tau}_\Delta^B > t]$$

$$+ \frac{1}{2}\tilde{P}(\tilde{\tau}_\Delta^B \le t)$$

$$= \tilde{P}(\tilde{\tau}_0^B \le t) + \frac{1}{2}\tilde{P}(\tilde{\tau}^B > t) + \frac{1}{2}\tilde{P}(\tilde{\tau}_\Delta^B \le t) .$$

Let $t \to \infty$ and do some algebra to finish the proof. □

(1.9) Corollary. Let $\{(\eta_t^A)\}$ be cancellative. If

(1.10) $$\lim_{t\to\infty} \sum_{\Lambda \neq \emptyset} |\tilde{P}(\tilde{\eta}_t^B = (\Lambda, 0)) - P(\tilde{\eta}_t^B = (\Lambda, 1))| = 0$$

for each $B \in S_0$, then $\{(\eta_t^A)\}$ is ergodic. In particular, ergodicity holds if

(1.11) $$\tilde{P}(\tilde{\tau}^B < \infty) = 1 \qquad \forall B \in S_0 .$$

Proof: The duality equation can be rewritten as

$$\psi_t^A(B) = \tilde{P}(\tilde{\tau}_0^B \le t) + \frac{1}{2}\tilde{P}(\tilde{\tau}^B > t) + \frac{1}{2}\tilde{P}(\tilde{\tau}_\Delta^B \le t)$$

$$+ \frac{1}{2} \sum_{\substack{\Lambda \ne \emptyset \\ \varepsilon = 0,1}} [\tilde{P}(\tilde{\eta}_t^B = (\Lambda, \varepsilon), |A \cap \Lambda| + \varepsilon \text{ even})$$

$$- \tilde{P}(\tilde{\eta}_t^B = (\Lambda, \varepsilon), |A \cap \Lambda| + \varepsilon \text{ odd})] .$$

Assuming (1.10), the last sum tends to 0 as $t \to \infty$. Thus $\lim_{t \to \infty} \psi_t^A(B)$ exists for each B, and is independent of A. This proves ergodicity. Finally, note that condition (1.11) implies (1.10). □

(1.12) Notes. Our treatment of cancellative systems generalizes the theory of spi systems with "$\alpha = 0$ duals," due to Holley and Stroock (1976a). In particular, Corollary (1.9) is based on one of their results. The graphical approach to cancellative systems is new.

2. Extralineal cancellative systems with pure births.

In contrast to the additive case, translation invariant extralineal cancellative systems need not be ergodic. One needs the presence of pure births to send the dual to Δ. The analogue of Theorem (II . 2.2) in the present setting is

(2.1) Theorem. Let $\{(\eta_t^A)\}$ be cancellative. If

(2.2)
$$\inf_{\Lambda \in S_0} \sum_{\substack{\text{pure birth } (i,z): \\ |\Lambda \cap V_{i,z}| \text{ odd}}} \lambda_{i,z} = \kappa > 0 ,$$

then the system is exponentially ergodic. In fact, (II . 2.1) holds with $\alpha = 2\kappa$.

Proof. Condition (2.2) ensures that the dual process $(\tilde{\eta}_t^B)$ goes to Δ with rate at least 2κ from any state other than $(\emptyset, 0)$, $(\emptyset, 1)$ and Δ. Thus $\tilde{\tau}^B < \infty$ P - a.s., and the duality equation yields

$$|\psi_t^A(B) - \psi^\nu(B)| = |\psi_t^A(B) - \tilde{P}(\tilde{\tau}_0^B < \infty) - \tfrac{1}{2}\tilde{P}(\tilde{\tau}_\Delta^B < \infty)|$$

$$\leq \tilde{P}(t < \tilde{\tau}_0^B < \infty) + \tfrac{1}{2}\tilde{P}(t < \tilde{\tau}_\Delta^B < \infty) + \tilde{P}(\tilde{\tau}^B > t)$$

$$\leq \tfrac{5}{2}\tilde{P}(\tilde{\tau}^B > t) \leq \tfrac{5}{2} e^{-2\kappa t} .$$

The rest is routine. □

(2.3) Corollary. Any translation invariant cancellative system with pure births is exponentially ergodic.

Proof. Pick $i \in I_0 : V_{i,0} \neq \emptyset$, $W_{i,0}(y) = \{y\} \; \forall y$. For any $\Lambda \in S_0$ we can find a $z \in Z^d$ such that $|\Lambda \cap V_{i,z}| = |\Lambda \cap (z + V_{i,0})| = 1$. Thus $\lambda \geq \lambda_{i,0}$ in (2.2). □

We now determine the general form of the flip rates for cancellative spin systems. One can assume that there is a single pure birth mechanism at each site x, say $V_{0,x} = \{x\}$, occurring at rate $\kappa_x \geq 0$. As in the additive case, for $i \geq 1$ we can take $W_{i,x}$ to be of the form

$$
\begin{aligned}
W_{i,x}(z) &= \{x, z\} && z \in C_{i,x} \\
&= \{z\} && z \notin C_{i,x}, \quad z \neq x \\
&= \{\emptyset\} && z = x \notin C_{i,x}
\end{aligned}
$$

for prescribed sets $C_{i,x} \in S_0$. But now, to get the most general cancellative spin system one must allow both $V_{i,x} = \emptyset$ and $V_{i,x} = \{x\}$, $i \geq 1$. As before, it is convenient to redefine I_x by removing $i = 0$. Then, denoting $I_x^0 = \{i \in I_x : V_{i,x} = \emptyset\}$, $I_x^1 = \{i \in I_x : V_{i,x} = \{x\}\}$, the flip rates for $\{(\eta_t^A)\}$ so induced have the form

$$(2.4) \qquad c_x(A) = \kappa_x + \sum_{\substack{i \in I_x^0 : \\ |(A \cap C_{i,x}) \,\Delta\, \{x\}| \text{ odd}}} \lambda_{i,x} + \sum_{\substack{i \in I_x^1 : \\ |(A \cap C_{i,x}) \,\Delta\, \{x\}| \text{ even}}} \lambda_{i,x} .$$

To ensure that there is no influence from ∞ , let

$$\lambda_x = \sum_{i \in I_x} \lambda_{i,x} ,$$

assume

(2.5) $$\sup_x \lambda_{i,x} < \infty ,$$

and also (II.1.6). Then Theorem (2.1) applies if

(2.6) $$\inf_x \kappa_x > 0 .$$

The duals $(\tilde{\eta}_t^B)$ for these spin systems may be thought of as <u>annihilating branching processes with parity</u>. With rate $\lambda_{i,x}$, a particle at $x \in \hat{\eta}_t^B$ tries to replace itself with particles located at $C_{i,x}$. At each site of $C_{i,x}$ which is already occupied, "annihilation" takes place, so the site becomes unoccupied. This describes the evolution of $(\hat{\eta}_t^B)$. (ε_t^B) simply flips back and forth between 0 and 1, a flip occurring each time a clock indexed by $i \in I_x^1$ effects $x \in \hat{\eta}_t^B$. Finally, a particle at $x \in \hat{\eta}_t^B$ sends the entire process to Δ at rate $2\kappa_x$.

(2.7) <u>Problem</u>. Show that one can assume $C_{i,x} \neq \{x\}$ $\forall i$, and $C_{i,x} \neq C_{j,x}$ if $i \in I_x^0$, $j \in I_x^1$, in the general representation (2.4) of cancellative spin systems.

(2.8) <u>Problem</u>. Derive the following results for cancellative systems by modifying the additive versions proved previously.

(i) If $\mu \in \mathfrak{M}$ is spatially mixing and $\mathcal{P}$ is local, then μP^t is spatially mixing for each $t < \infty$.

(ii) If (2.2) holds and $\mathcal{P}$ is local and translation invariant, then the unique equilibrium ν for $\{(\eta_t^A)\}$ has exponentially decaying correlations.

(2.9) <u>Notes</u>. For another approach to the theory of cancellative spin systems see Holley and Stroock (1976d). Theorem (2.1) generalizes a result from Holley and Stroock (1976a).

3. Application to the stochastic Ising model.

The basic d-dimensional stochastic Ising model is the spin system on Z^d with flip rates

$$(3.1)\qquad c_x(A) = [1 + \exp\{-\theta U_x(A)\}]^{-1} ,$$

where

$$U_x(A) = 4(2A(x) - 1)(d - |A \cap N_x|)$$

($N_x = \{y \in Z^d : |y-x| = 1\}$). $\theta \geq 0$ is a parameter. This system $\{(\eta_t^A)\}$ is one of the simplest and most widely studied models for the evolution of a physical system with two possible states per site (e.g. solid or liquid, "spin up" or "spin down" in a piece of iron, etc.). Background and motivation for the choice (3.1) will be found in the papers mentioned in the Notes (3.3). The construction of $\{(\eta_t^A)\}$ for arbitrary values of θ requires methods which will not be discussed here. For certain parameter values, however, the stochastic Ising model has a cancellative representation, and in these cases our methods apply. The systems with flip rates (3.1) are important because the Gibbs measures with potential U are equilibria for them. $\mu \in \mathfrak{M}$ is such a measure if μ has positive cylinders and

$$(3.2)\qquad \mu([A,\{x\}] \mid [A,\Lambda]) = [1 + \exp\{\theta U_x(A)\}]^{-1} ,$$

where $[A,\Lambda] = \{B \in S : B \cap \Lambda = A \cap \Lambda\}$, for all finite $\Lambda : N_x \subset \Lambda \subset Z^d - \{x\}$. Given such a μ, if we write ${}_xA = A \Delta x$ for the configuration "A flipped at x" and put $\overline{\Lambda} = \Lambda \cup x$, then for Λ as above,

$$\begin{aligned}\mu([A,\overline{\Lambda}])c_x(A) &= \mu([A,\overline{\Lambda}])\,\mu([{}_xA,\{x\}] \mid [A,\Lambda]) \\ &= \mu([{}_xA,\overline{\Lambda}])\,\mu([A,\{x\}] \mid [{}_xA,\Lambda]) \\ &= \mu([{}_xA,\overline{\Lambda}])\,c_x({}_xA) .\end{aligned}$$

Roughly, then, the flow from 0 to 1 equals the flow from 1 to 0 at each site when the stochastic Ising model is started in μ. This suggests that the

Gibbs measure μ is invariant for $\{(\eta_t^A)\}$, a fact which can be proved rigorously. It turns out that the stationary process starting from μ is <u>time reversible</u>, i.e. has the same dynamics whether time runs backwards or forwards.

Suppose now that $d = 1$. Then the stochastic Ising model is cancellative for every $\theta \geq 0$. Indeed, it is simply a voter model with pure births. To see this, take $\kappa_x \equiv (1 + e^{4\theta})^{-1}$, $I_x^0 = \{1,2\}$, $I_x^- = \emptyset$, $C_{1,x} = \{x-1\}$, $C_{2,x} = \{x+1\}$, $\lambda_{1,x} = \lambda_{2,x} = \frac{1}{2} - (1 + e^{4\theta})^{-1}$. Then (2.4) coincides with (3.1). Moreover, since $\kappa = (1 + e^{4\theta})^{-1} > 0$, $\{(\eta_t^A)\}$ is exponentially ergodic. In particular, there is only one equilibrium for the model, so there is only one Gibbs state ν with potential U. By Problem (2.8 ii), ν has exponentially decaying correlations. In the language of statistical physics, there is no phase transition when $d = 1$.

The situation when $d = 2$ is much more interesting. A famous result of Onsager asserts that there is more than one Gibbs measure μ with potential U if and only if $\theta > \theta_* = \text{arc sinh } 1 \approx .88$. For $\theta > \theta_*$, the stochastic Ising model is therefore nonergodic. This is one of the simplest examples of a translation invariant local spin system with strictly positive flip rates which is nonergodic. It is not known whether such a system exists in one dimension. A straightforward computation shows that the 2-dimensional stochastic Ising model has a cancellative representation if and only if $\theta \leq \frac{\ln 3}{4} \approx .27$. For θ in this range, the mechanisms affecting each site $\cdot \in Z^2$ may be described pictorially as follows:

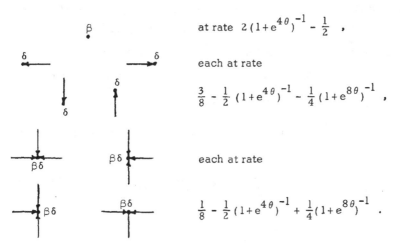

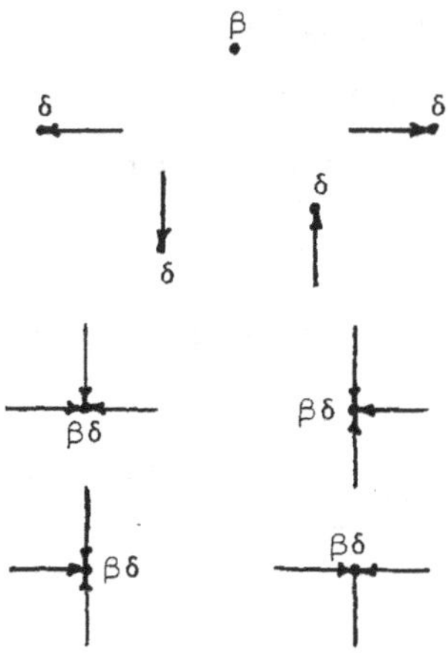

at rate $2(1+e^{4\theta})^{-1} - \frac{1}{2}$,

each at rate

$\frac{3}{8} - \frac{1}{2}(1+e^{4\theta})^{-1} - \frac{1}{4}(1+e^{8\theta})^{-1}$,

each at rate

$\frac{1}{8} - \frac{1}{2}(1+e^{4\theta})^{-1} + \frac{1}{4}(1+e^{8\theta})^{-1}$.

A routine check shows that these rates give rise to the flip rates (3.1) when $d = 2$. While the last two rates of the substructure are always non-negative, the rate of pure birth is only non-negative if $\theta \le \frac{\ell n\, 3}{4}$. When $\theta < \frac{\ell n\, 3}{4}$ we conclude, just as in one dimension, that $\{(\eta_t^A)\}$ is exponentially ergodic, that there is a unique Gibbs measure ν with potential U, and that ν has exponentially decaying correlations.

(3.3) Notes. A beautiful introduction to the Ising model is Griffiths (1972); for a nice treatment of more general Gibbs and Markov random fields, see Preston (1974). Stochastic Ising models were first studied by Glauber (1963). More recent applications of the dynamical models to the equilibrium theory may be found in Holley (1974) and Holley and Stroock (1976b, 1976c). The connections between time-reversible spin systems and Gibbs random fields are discussed by Logan (1974). The results of this section dealing with the 2-dimensional Ising model were obtained by Stavskaya (1975) and Holley and Stroock (1976a, 1976b).

4. Generalized voter models.

If the substructure which gives rise to $\{(\eta_t^A)\}$ has no pure births, then the cancellative duality equation simplifies to

$$P(|\eta_t^A \cap B| \text{ even}) = P(|\hat{\eta}_t^B \cap A| + \varepsilon_t^B \text{ even}) .$$

For spin systems of this type there is a cancellative analogue of Theorem (II.3.2). Both the statement and the proof are virtually identical: the hypotheses imply that $(|\hat{\eta}_t^B|)$ is dominated by a critical or subcritical Galton-Watson process, so that $(\tilde{\eta}_t^B)$ hits $(\emptyset, 0)$ or $(\emptyset, 1)$ eventually with probability one. We state the result formally, but omit the proof.

(4.1) Theorem. Let $\{(\eta_t^A)\}$ be a cancellative spin system having flip rates (2.4) with $\kappa_x \equiv 0$. Set

$$\iota = \inf_{x \in Z^d} \lambda_x , \quad m = \sup_{\substack{x \in Z^d: \\ \lambda_x > 0}} \sum_{i \in I_x} \frac{\lambda_{i,x}}{\lambda_x} |C_{i,x}| .$$

If $\iota > 0$ and $m < 1$, then the system is exponentially ergodic. In the translation invariant case the system is also ergodic if $m = 1$ and $C_{i,x} = C_i = \emptyset$ for some i.

A particularly simple family of systems without pure births to which Theorem (4.1) does not apply consists of the (translation invariant) generalized voter models, where $|C_{i,x}| = 1$ for all i. As in the additive case, it is convenient to take $I_x = Z^d$. After some manipulation, the flip rates for generalized voter models can be written in the form

$$(4.2) \qquad c_x(A) = \frac{\lambda}{2}(1 + (1-2A(x))[\sum_{z \in (x+I^0) \cap A} p_{z-x} - \sum_{z \in (x+I^1) \cap A} p_{z-x}])$$

for some $\lambda > 0$, probability density $p = (p_z; z \in Z^d)$, and disjoint subsets I^0 and I^1 of Z^d such that $I^0 \cup I^1 = \text{support } p = \{z : p_z > 0\}$. By a constant change of time scale, one can assume that $\lambda = 1$.

(4.3) Problems. Show how to get (4.2) from (2.4). Show also that any of the voter models of section (II.7) has a cancellative representation of the form (4.2) with $I^1 = \emptyset$. What other systems have both additive and cancellative representations?

The generalized voter models for which $I^0 = \emptyset$ are the anti-voter models of Problem (1.3). When $I^0 \neq \emptyset$ and $I^1 \neq \emptyset$ we can think of the model as having a "voter component" and an "anti-voter component." For the basic anti-voter model on Z^1, where p is the simple random walk transition density, it is clear that $\{(\eta_t^A)\}$ is not ergodic. This is because the configurations

$$A_0 = \{\text{even integers}\} \quad \text{and} \quad A_1 = \{\text{odd integers}\}$$

are both traps. If we modify the model by taking $p_{-1} = p_2 = \frac{1}{2}$, for example, then the resulting $\{(\eta_t^A)\}$ does not have traps, so it seems plausible that ergodicity should result. We now prove this, as an application of Corollary (3.4). In fact, we give a necessary and sufficient condition for a generalized voter model with flip rates (4.2) to be ergodic. To avoid trivialities, we assume that p is irreducible, i.e. the group spanned by the support of p is all of Z^d.

(4.4) Theorem. Let $\{(\eta_t^A)\}$ be cancellative, with flip rates of the form (4.2) for some irreducible density p. The system is ergodic if there are integers m_z, $z \in I^0 \cup I^1$, only finitely many non-zero, such that

$$\sum m_z \cdot z = 0 \quad \text{and} \quad \sum_{z \in I^1} m_z \text{ is odd .} \tag{4.5}$$

Otherwise the system is nonergodic.

Note that the system having $p_{-1} = p_2 = \frac{1}{2}$ is ergodic, since we can apply the theorem with $m_{-1} = 2$, $m_2 = 1$.

Proof. The dual processes $(\tilde{\eta}_t^B)$ for generalized voter models are annihilating random walks with parity; we will verify (1.10) for these processes. The argument is based on comparison of $(\tilde{\eta}_t^B)$ with processes $({}_s\tilde{\eta}_t^B) = ({}_s\hat{\eta}_t^B, {}_s\varepsilon_t^B)$, $s \geq 0$, which ignore the annihilation rule after time s. Naturally we must enlarge the state space and basic probability space to allow for multiple (but finite) occupancy of sites. Using a more elaborate graphical representation, this can be done in such a way that

$$ {}_s\tilde{\eta}_t^B = \tilde{\eta}_t^B \qquad \text{for} \qquad t \leq s , \tag{4.6}$$

$$\hat{\eta}_t^B \text{ behaves like independent random walks after time } s . \tag{4.7}$$

Let $\check{S}$ be the extended state space for the $({}_s\hat{\eta}_t^B)$. Write $p_{B\Lambda}^0(t) = \tilde{P}(\tilde{\eta}_t^B = (\Lambda, 0))$, $p_{B\Lambda}^1(t) = \tilde{P}(\tilde{\eta}_t^B = (\Lambda, 1))$. Define ${}_sp_{B\Lambda}^0(t)$ and ${}_sp_{B\Lambda}^1$ analogously in terms of ${}_s\tilde{\eta}_t^B$, where Λ is a generic element of $\check{S}$. According to (1.10) we want to show that

$$\lim_{t \to \infty} \sum_{\emptyset \neq \Lambda \in S_0} |p_{B\Lambda}^0(t) - p_{B\Lambda}^1(t)| = 0 \qquad \forall B \in S_0 . \tag{4.8}$$

The sum may be extended to $\Lambda \in \check{S}$, and majorized by

$$\sum_{\emptyset \neq \Lambda \in \check{S}} |p^0_{B\Lambda}(t) - {}_s p^0_{B\Lambda}(t)| + \sum_{\emptyset \neq \Lambda \in \check{S}} |p^1_{B\Lambda}(t) - {}_s p^1_{B\Lambda}(t)|$$

$$+ \sum_{\emptyset \neq \Lambda \in \check{S}} |{}_s p^0_{B\Lambda}(t) - {}_s p^1_{B\Lambda}(t)|$$

$$= \sum_1^s(t) + \sum_2^s(t) + \sum_3^s(t) \ .$$

To estimate the first two sums we use the "fundamental coupling inequality": if X_1 and X_2 are (general) random variables governed by a joint probability measure P, and if μ_1 and μ_2 are their respective laws, then $\|\mu_1 - \mu_2\| \le P(X_1 \neq X_2)$. Using (4.5), the conclusion is that for $t > s$, $\sum_1^s(t) + \sum_2^s(t) \le 2 \cdot P$(the dual has a collision between times s and t) $\le 2 \cdot P$ (the dual has a collision after s). Since $(\hat{\eta}_t^B)$ has finitely many particles two of which disappear with each collision, $\lim_{s\to\infty} \sup_{t \ge s} (\sum_1^s(t) + \sum_2^s(t)) = 0$. Next, apply the Markov property at time s to $\sum_3^s(t)$. It follows that (1.10) holds if

$$\lim_{t\to\infty} \sum_{\emptyset \neq \Lambda \in \check{S}} |\, {}_0 p^0_{B\Lambda}(t) - {}_0 p^1_{B\Lambda}(t) \,| = 0 \qquad B \in \check{S}, \tag{4.9}$$

i.e. if the analogue of (4.8) holds for the totally independent process. Clearly $|{}_0\hat{\eta}_t^B| = |B|$ for all t. Also,

$${}_0 p^0_{B\Lambda}(t) = \check{P}(\{{}_0\hat{\eta}_t^B = \Lambda\} \cap E_0)$$

and

$${}_0 p^1_{B\Lambda}(t) = \check{P}(\{{}_0\hat{\eta}_t^B = \Lambda\} \cap E_1),$$

where E_0 (and E_1) are respectively the events that the total number of displacements from I^1 through time t by the $|B|$ independent random walks is even (and odd). Using these observations and (4.7), it is not hard to see that the sum in (4.9) is majorized by

(4.10) $$|B|! \sum_{z \in Z^d} | p^0_{0z}(t) - p^1_{0z}(t) | .$$

Consider $\mathbb{A} = Z^d \times \{0,1\}$ as an additive group with addition mod 2 in the second coordinate. The one particle dual performs a random walk on $\mathbb{A}$. Denoting the walk which starts at $(x,\varepsilon) \in A$ by $(X_t^{(x,\varepsilon)})$, we note that $p^0_{0z}(t) = \Pr(X_t^{(0,0)} = z)$ and $p^1_{0z}(t) = \Pr(X_t^{(0,1)} = z)$. Let $\tilde{I} = \{(i,0) : i \in I^0\} \cup \{(i,1) : i \in I^1\}$, and let $\mathbb{B}$ be the subgroup of $\mathbb{A}$ generated by $\tilde{I}$. The hypothesis (4.5) is equivalent to $(0,1) \in \mathbb{B}$. Now it is well known that

(4.11) $$\| \Pr(X_t^x \in \cdot) - \Pr(X_t^y \in \cdot) \| \to 0 \quad \text{as} \quad t \to \infty$$

for any x, y in the same irreducible set of states of a continuous time random walk on a countable abelian group. But $(0,1) \in \mathbb{B}$ says that $(0,0)$ and $(0,1)$ communicate in $\mathbb{A}$, so (4.11) holds for $x = (0,0)$, $y = (0,1)$. Thus the variation norm in (4.10) tends to 0 as $t \to \infty$ for each $B \in S_0$. This completes the proof of ergodicity when (4.5) holds. If (4.5) does not hold, i.e. if $(0,1) \notin \mathbb{B}$, then any $b \in \mathbb{B}$ has a unique representation $b = (z, \varepsilon_z)$. Define $A_0 = \{z : \varepsilon_z = 0\}$, $A_1 = \{z : \varepsilon_z = 1\}$. It is easy to check that A_0 and A_1 are traps for $\{(\eta_t^A)\}$, so the system is nonergodic. The proof is finished. □

(4.12) Problem. Show that an anti-voter model is ergodic if and only if its density p has odd period in Z^d, i.e. $Z^d/\text{grp}(I^1 - I^1) =$ an odd positive integer.

(4.13) Notes. This section is adapted from Griffeath (1977). Our approach to generalized voter models is based on the treatment of anti-voter models of Holley and Stroock (1976d). Anti-voter models were first studied by Matloff (1977).

5. Annihilating random walks.

For lineal cancellative systems $\{(\eta_t^A)\}$ the second coordinate processes (ε_t^B) of the duals $(\tilde{\eta}_t^B)$ are identically 0, so we can discard them. $\tilde{P}$ reduces to $\hat{P}$, and we get the symmetric duality equation

$$P(|\eta_t^A \cap B| \text{ even}) = \hat{P}(|\hat{\eta}_t^B \cap A| \text{ even}) \qquad A \in S, \quad B \in S_0 .$$

Just as in the additive lineal setting, the dual can be extended to a Markov family on all of S. We state the analogue of Theorem (II.3.1), but omit the easy proof.

(5.1) Theorem. Let $\{(\eta_t^A)\}$ be the cancellative system induced by a lineal substructure $\mathcal{P}$, $\{(\hat{\eta}_t^B); B \in S\}$ the lineal cancellative system induced by the dual substructure $\hat{\mathcal{P}}$. For each $t > 0$, $A, B \in S$, at least one finite,

$$\psi_t^A(B) = \hat{\psi}_t^B(A) . \tag{5.2}$$

There is an invariant measure $\nu \in \mathfrak{M}$ such that $\mu_{\frac{1}{2}} P^t \rightarrow \nu$ as $t \rightarrow \infty$. Moreover,

$$\{(\eta_t^A)\} \text{ ergodic} \iff \nu = \delta_\emptyset$$

$$\iff \hat{P}(\hat{\tau}_\emptyset^A < \infty) = 1 \qquad \forall A \in S_0 .$$

The remainder of this section will be devoted to one of the simplest types of lineal cancellative system: the annihilating random walks. These are defined as in Example (I.1.3), except that the transition density $p = (p_z; z \in Z^d)$ can be arbitrary. Alternatively, they are the extended dual systems for voter models in their cancellative representations. They share many features with the coalescing random walks of Section (II.9), but interesting and sometimes surprising differences arise.

Let us first prove ergodicity. Just as for the coalescing random walks, one expects collisions to yield a limiting measure of $\delta_\emptyset$ starting from any $\mu \in \mathfrak{M}$. In fact, the two types of random walks with interference for given p can clearly be defined on the same substructure $\mathcal{P}$ in such a way that $\eta_t^A \subset \xi_t^A$ for all t. Ergodicity of $\{(\eta_t^A)\}$ then follows from that of $\{(\xi_t^A)\}$ (Proposition (II.9.1)):

(5.3) Proposition. Let $\{(\eta_t^A)\}$ be the annihilating random walks with transition density p. Then

$$\mu P^t \rightarrow \delta_\emptyset \quad \text{as} \quad t \rightarrow \infty \qquad \forall \mu \in \mathfrak{M} .$$

Since $\eta_t^A \subset \xi_t^A$, and the distribution of the latter tends to $\delta_\emptyset$, the quantities

$$\frac{P(\Lambda \subset \eta_t^A)}{P(\Lambda \subset \xi_t^A)}$$

are of interest for large t. Here we compute the limit only in one very special case. Arratia (private communication) has recently proved that the above ratio tends to $2^{-|\Lambda|}$ for the basic models in any dimension d.

(5.4) Proposition. Let $\{(\xi_t^A)\}$ and $\{(\eta_t^A)\}$ be basic coalescing and annihilating random walks respectively on Z. Then

$$\lim_{t\to\infty} \frac{P(0 \in \eta_t^Z)}{P(0 \in \xi_t^Z)} = \frac{1}{2} .$$

Proof. In any dimension d, and for arbitrary transition function p,

$$\frac{P(0 \in \eta_t^{Z^d})}{P(0 \in \xi_t^{Z^d})} = \frac{P(|\zeta_t^0| \text{ odd})}{P(\zeta_t^0 \neq \emptyset)} ,$$

where (ζ_t^0) is the corresponding voter process starting at $\{0\}$. It seems intuitively clear that this ratio should always tend to $\frac{1}{2}$, a fact which has been proved by R. Arratia for $d > 1$. When $d = 1$ and $p_{-1} = p_1 = \frac{1}{2}$, the reflection principle yields

$$P(\zeta_t^0 = \emptyset) \sim \frac{1}{\sqrt{\pi t}} , \quad P(|\zeta_t^0| \text{ odd}) \sim \frac{1}{2\sqrt{\pi t}} \quad \text{as} \quad t\to\infty$$

(cf. Theorem (II . 7.11)), which proves the claim. □

(5.5) Problem. For the basic one dimensional annihilating and coalescing random walks, compute

$$\lim_{t\to\infty} \frac{P(\{0,1\} \subset \eta_t^Z)}{P(\{0,1\} \subset \xi_t^Z)} .$$

Next we derive asymptotics for the mean interparticle distance in the basic annihilating random walks on Z. The result is similar to one of the Problems (II . 9.7), but a curious distinction arises.

(5.6) Theorem. Let $\mu = \mu_f$ be a renewal measure (defined as in the proof of Theorem (II.4.7)) such that the distribution of $f*f$ is aperiodic. Write $D(A)$ for the mean interparticle distance in A (provided it is well-defined). If (η_t^μ) is the basic one-dimensional annihilating random walks starting from μ, then

$$(5.7)\qquad \lim_{t\to\infty} \frac{D(\eta_t^\mu)}{\sqrt{t}} = 2\sqrt{\pi} \quad \text{in P-probability.}$$

In particular, (5.7) holds for any product measure μ_θ, $\theta \in (0,1)$. However, there are mixing translation invariant measures μ such that (5.7) does not hold

Proof. Using the same method as in Theorem (II.7.11), one first shows that if μ is translation invariant mixing, $\mu \neq \delta_\emptyset$, then

$$D(\eta_t^\mu) = [P(0 \in \eta_t^\mu)]^{-1} \qquad P\text{ - a.s.}$$

The integrated form of (5.2) is

$$\psi_t^\mu(B) = \hat{E}[\psi^\mu(\hat{\eta}_t^B)] \quad .$$

Manipulating this we get

$$\frac{1}{2}P(\hat{\eta}_t^0 \neq \emptyset) - P(0 \in \eta_t^\mu)$$
$$= \hat{E}[\psi^\mu(\hat{\eta}_t^0) - \frac{1}{2}, \ \hat{\eta}_t^0 \neq \emptyset] \ .$$

Choose $g(t) = o(t^{\frac{1}{4}})$, such that $g(t) \to \infty$ as $t \to \infty$, and estimate

$$\sqrt{t}\,\hat{E}[|\psi^\mu(\hat{\eta}_t^0) - \tfrac{1}{2}|, \ \hat{\eta}_t^0 \neq \emptyset]$$
$$(5.8)\qquad \leq \sqrt{t}\,\hat{P}(0 < |\hat{\eta}_t^0| \leq g(t))$$
$$+ \sup_{[x,y]:\, y-x > g(t)} |\psi^\mu([x,y]) - \tfrac{1}{2}|\sqrt{t}\,\hat{P}(\hat{\eta}_t^0 \neq \emptyset) \ .$$

Recall that $(|\hat{\eta}_t^0|)$ is a rate 2 simple random walk starting at 1 and with absorption at 0. We have already noted that $\sqrt{t}\ P(\hat{\eta}_t^0 \neq \emptyset) \to \sqrt{\pi}^{-1}$ as $t \to \infty$, while the reflection principle and central limit theorem imply that the first term on the right side of (5.8) tends to 0 as $t \to \infty$. To conclude that

$$\lim_{t\to\infty} \sqrt{t}\, P(0 \in \eta_t^{\mu}) = \frac{1}{2} \lim_{t\to\infty} \sqrt{t}\, P(\hat{\eta}_t^{0} \neq \emptyset) = \frac{1}{2\sqrt{\pi}} ,$$

it suffices to check that

(5.9) $$\lim_{n\to\infty} \psi^{\mu}([0,n]) = \frac{1}{2} .$$

Let ω be μ-distributed, and write $N_{\Lambda}(\omega) = |\omega \cap \Lambda|$ = the number of renewals on $\Lambda \in S_0$. Define

$$p_n = \mu\{N_{[1,n-1]} \text{ even} \mid \omega(0) = 1\} , \qquad n \geq 1 ,$$

and note that (p_n) satisfies the renewal equation

$$p_n = a_n + \sum_{k=1}^{n-1} (f * f)_k\, p_{n-k} ,$$

where $a_n = \sum_{k=n}^{\infty} f_n$. By hypothesis $f * f$ is aperiodic, and so the renewal theorem asserts that

$$p_n \to \frac{\sum_{k=1}^{\infty} \sum_{\ell=k}^{\infty} f_{\ell}}{\sum_{k=1}^{\infty} k\cdot(f*f)_k} = \frac{m}{2m} = \frac{1}{2} \quad \text{as } n\to\infty .$$

The rest of the proof of (5.9) is routine. By working harder one can dispense with the aperiodicity condition, and show (5.7) for any renewal μ. To get an example of a mixing translation invariant μ such that (5.9) does not hold, start with μ_θ, $\theta \in (0,1) - \{\frac{1}{2}\}$, and consider the "border measure" $\tilde{\mu}_\theta$ defined by

$$\varphi^{\tilde{\mu}_\theta}(\Lambda) = \mu_\theta\{A : A(x-1) = A(x) ,\ x \in \Lambda\} .$$

There is a "border equation" which connects $\tilde{\eta}_t^{\mu}$ and $\hat{\eta}_t^{\mu}$. It is based on the observation that the "borders" between voters of opposite opinion execute annihilating random walks. Thus, for any $\mu \in \mathfrak{M}$ and its corresponding $\tilde{\mu} \in \mathfrak{M}$, $x, y \in Z$, $t \geq 0$,

$$P(|\tilde{\eta}_t^{\mu} \cap [x+1, y]| \text{ even}) = \hat{P}(\hat{\eta}_t^{\mu}(x) = \hat{\eta}_t^{\mu}(y)) .$$

Using this equation and Theorem (II . 7.11), it is easy to see that

$$\lim_{t\to\infty} E\left[\frac{D(\eta_t^{\tilde{\mu}_\theta})}{\sqrt{t}}\right] = \lim_{t\to\infty} E\left[\frac{D(\eta_t^{\hat{\mu}_\theta})}{\sqrt{t}}\right] = \frac{\sqrt{\pi}}{2\theta(1-\theta)} \neq 2\sqrt{\pi}$$

if $\theta \in (0,1)$, $\theta \neq \frac{1}{2}$. The proof is finished. □

Something like the renewal assumption is apparently necessary in Theorem (5.6) in order to guarantee a certain uniformity of the spacing of the particles. The result is especially interesting because it gives an example of limiting behavior which is insensitive to density but feels some of the other structure of the initial μ.

To conclude our discussion of annihilating random walks, we consider the question of "site recurrence": is the origin occupied at arbitrarily large times? The easy Problem (II. 9.5) asserts that the answer is yes for irreducible recurrent <u>coalescing</u> random walks whenever $A \neq \emptyset$. For annihilating (η_t^A), the situation is altogether different. To see this, let $d = 1$, choose p irreducible recurrent, and let A be an initial configuration of the form $A = \{1, x_1, x_1+1, x_2, x_2+1, \cdots\}$, where $1 < x_1 < x_1+1 < x_2 < x_2+1 < \cdots$. By letting $x_{i+1} - x_i$ tend to ∞ sufficiently fast, one can ensure that for all sufficiently large i the particles starting at x_i and x_i+1 annihilate one another before either has a chance to collide with any other particle. The remaining odd number of particles will reduce to one eventually, which will perform a recurrent random walk from then on. Thus every site of Z will be visited infinitely often. By adding a <u>single</u> particle at 0 to A, the recurrence property is changed completely. Now every point of Z is visited only finitely often. The annihilating case is evidently much more delicate than the coalescing case.

(5.10) <u>Problem</u>. Let $\{(\xi_t^A)\}$ be the basic one-dimensional voter model. Find infinite configurations A, B such that

$$\frac{1}{t}\int_0^t f(\xi_s^A)\,ds \to f(\emptyset) \qquad P\text{-a.s.} \qquad \forall f \in C,$$

$$P\left(\frac{1}{t}\int_0^t f(\xi_s^B)\,ds \text{ fails to converge}\right) = 1 \quad \text{for some} \quad f \in C.$$

(Hint: one approach is to use the "border equation" and the discussion of the last paragraph.) Thus, in contrast to the biased voter model, the pointwise ergodic behavior of the unbiased voter model is unstable.

Starting from "nice" initial states A, the annihilating random walks (η_t^A) with any transition density p in any dimension d are point recurrent. To keep matters simple, let us consider $A = Z^d$, in which case a theorem corresponding to (II.9.2) can be proved.

(5.11) Theorem. Let $\{(\eta_t^A)\}$ be annihilating random walks on Z^d with density p. Then

$$P(\limsup_{t \to \infty} \eta_t^{Z^d}(0) = 1) = 1 .$$

The proof is somewhat involved, so we will only sketch it. The idea is to proceed by comparison with the coalescing case. In Theorem (II.9.2) it was shown that the expected amount of time that the voter process (ζ_t^0) lives before absorption is infinite. It follows that the expected amount of time (ζ_t^0) spends in states of odd cardinality before absorption is also infinite, and this is precisely the expected amount of time that $(\eta_t^{Z^d})$ occupies 0. However the estimates on $P(\tau_t \in [t,u])$ which were used for the coalescing walks only yield

$$P(\limsup_{t \to \infty} \eta_t^{Z^d}(0) = 1) > 0$$

Thus we need a 0 - 1 law to finish the proof. One approach is to show that $(\eta_t^{Z^d})$ is spatially mixing, in the sense that

$$\lim_{|y| \to \infty} P(F \cap \theta^y G) = P(F)\, P(G) ,$$

where θ^y is the shift $\theta^y(x) = x + y$, and F and G are any events measurable with respect to the process $(\eta_t^{Z^d})$. Taking $E_x = \{\limsup_{t \to \infty} \eta_t^{Z^d}(x) = 1\}$, $E = \bigcap_{x \in Z^d} E_x$, and noting that $E = \theta^y E$ for all y, we get $P(E) = 0$ or 1. The proof is completed by showing that if $P(E_0) > 0$, then $P(E) > 0$, and hence $P(E_0) = P(E) = 1$. Note that we make use of the translation invariance of δ_{Z^d} in this argument.

Another approach, which works when p is transient relies on the following

0 - 1 law.

(5.12) Lemma. Let $\{(\eta_t^A)\}$ be annihilating random walks on Z^d with a transient density p, and denote

$$E^A = \{\limsup_{t \to \infty} \eta_t^A(0) = 1\} .$$

Then

$$P(E^A) = 0 \ \underline{\text{or}}\ 1 \quad \text{for each} \quad A \in S ,$$

and

$$P(E^{A \cup B}) = P(E^A) \qquad A \in S, \quad B \in S_0 .$$

Proof. For each $x \in Z^d$, let (X_t^x) be a random walk with density p starting from x, all walks independent and governed by a joint probability law P. The processes (η_t^A) can be represented on this space in the obvious way. Note, however, that this is not the standard graphical representation: e.g. the processes $(\eta_t^{\{x\}}) = (X_t^x)$ are independent. We show that for any $x \in A$, $E^A \approx E^{A-x}$, i.e. E^A and E^{A-x} differ by a P-null set. It will follow that, up to $\approx$, E^A does not depend on the trajectories of any finite number of walks (X_t^x), and hence the Kolmogorov 0 - 1 law applies. This will also yield $P(E^{A \cup B}) = P(E^A)$ $A \in S$, $B \in S_0$. Fix a particular realization $(X_t^x; t \geq 0, z \in Z^d)$ of the independent walks. For $A \in S$, $y \in Z^d$, let n_A be the total number of arrivals at 0 by particles in (η_t^A). Thus $0 \leq n_A \leq \infty$. Now suppose $x_0 \in A$ is removed. This sets off a "chain of effects" which determine n_{A-x}. First of all, the visits of $(X_t^{x_0})$ to 0 before it collides with some other "living" walk $(X_t^{x_1})$ must be subtracted from n_A. Next, the visits of $(X_t^{x_1})$ to 0 from the collision time with $(X_t^{x_0})$ until the next collision with a living walk $(X_t^{x_2})$ must be added to n_A. This is because $(X_t^{x_0})$ no longer annihilates $(X_t^{x_1})$. The chain continues in this manner, with alternating positive and negative contributions, either indefinitely, or until a walk lives forever. If we consider the unique trajectory formed by those portions of the walks $(X_t^{x_i})$ which are involved in the chain of effects, one can check that as a

P-stochastic process it is simply a random walk governed by p. This walk is assumed transient, so it visits 0 only finitely often. In other words, the alternating positive and negative contributions to $n_A - n_{A-x}$ are all 0 from some time on with P-probability one. Hence n_{A-x} is infinite if and only if n_A is. This completes the proof.

(5.13) Problem. Assume that p is recurrent. With E^A defined as above, use the standard graphical representation, and Proposition (1.2) in particular, to show that

$$P(E^A) = P(E^{A \cup B})$$

whenever $A \in S$, $B \in S_0$, $A \cap B = \emptyset$ and $|B|$ is even.

(5.14) Problem. Let $(\xi_t^{\mu_{\frac{1}{2}}})$ be the basic voter process on Z starting from $\frac{1}{2}$ - $\frac{1}{2}$ product measure. Does the voter at the origin change opinion infinitely often?

(5.15) Notes. Annihilating random walks were first considered by Erdös and Ney (1974), who conjectured that the basic one dimensional process (η_t^{Z-0}) was site recurrent. Their conjecture was confirmed in various nearest neighbor settings by Adelman (1976), Lootgieter (1977) and Schwartz (1978). The recurrence question for general transition density p and dimension d is studied in Griffeath (1978b); the approach here is based on that paper. Holley and Stroock (1976d) identified the annihilating random walks $\{(\eta_t^A); A \in S_0\}$ as dual processes for voter models with respect to the "$\alpha = 0$" basis. Theorem (5.6) is taken from Bramson and Griffeath (1978b), where additional results on the "dispersion" of the basic one dimensional system may be found.

CHAPTER IV: UNIQUENESS AND NONUNIQUENESS

1. Additive and cancellative pregenerators.

This final chapter deals with the uniqueness problem for additive and cancellative systems:

When do the local dynamics of $\{(\xi_t^A)\}$ and $\{(\eta_t^A)\}$ uniquely determine the particle system?

We will concentrate on the additive case; the analogous results for cancellative systems will usually be left as exercises. To formulate uniqueness questions rigorously, one makes use of the generator of a Markov process; we will assume throughout this chapter that the reader is familiar with generators.

Let $\mathcal{P}$ be a percolation substructure, and consider the additive system induced by $\mathcal{P}$. Recall that when the i'th clock at x goes off at rate $\lambda_{i,x}$, configuration A is replaced by $\mathfrak{A}_{i,x}(A)$ (as defined in (I.2.2)). Remember also that $\mathfrak{F}_\Lambda$ denotes those $f: S \to R$ which depend only on sites in Λ, $\mathfrak{F} = \bigcup_{\Lambda \in S_0} \mathfrak{F}_\Lambda$, and $C = \{\text{continuous } f: S \to R\}$. Define the additive pregenerator $G: \mathfrak{F} \to C$ induced by $\mathcal{P}$ as

$$Gf(A) = \sum_{i,x} \lambda_{i,x} [f(\mathfrak{A}_{i,x}(A)) - f(A)] .$$

(Our assumptions on $\mathcal{P}$ ensure that $Gf \in C$ for each $f \in \mathfrak{F}$.) We say that there is a unique additive system with substructure $\mathcal{P}$ if there is a unique extension G^e of G which is the generator of a Markov process.

Similarly, consider the cancellative system induced by $\mathcal{P}$. When the i'th clock at x goes off at rate $\lambda_{i,x}$, configuration $A \in S$ is replaced by $\mathfrak{C}_{i,x}(B)$ (cf. the first paragraph of (III.1).) The cancellative pregenerator $G: \mathfrak{F} \to C$ induced by $\mathcal{P}$ is

$$Gf(A) = \sum_{i,x} \lambda_{i,x} [f(\mathfrak{C}_{i,x}(A)) - f(A)] .$$

(Again, our assumptions on $\mathcal{P}$ ensure that $Gf \in C$.) There is a unique cancellative system with substructure $\mathcal{P}$ if there is a unique extension G^e of G which is the generator of a Markov process.

In Chapters 2 and 3 we made the assumption

(1.1) $$P(\text{influence from } \infty) = 0$$

to ensure that $\{(\xi_t^A)\}$ and $\{(\eta_t^A)\}$ be Feller. If (1.1) does not hold, then genuine alternatives to the canonical definitions (II.1.1) and (III.1.1) arise. In the additive setting,

(1.2) $$\bar{\xi}_t^A = \{x : N_t^A(x) > 0 \text{ \underline{or} strong influence to } (x,t) \text{ from } \infty\}$$

and

(1.3) $$\bar{\bar{\xi}}_t^A = \{x : N_t^A(x) > 0 \text{ \underline{or} influence to } (x,t) \text{ from } \infty\}$$

also define systems with pregenerator G. The simplest nonuniqueness examples arise in this manner.

(1.4) Problem. Consider flip rates on Z of the form

$$c_x(A) = x^2 \, [A(x) + (1-2A(x))A(x+1)] \qquad x \geq 1$$

$$= 0 \qquad\qquad x \leq 0 \, .$$

Show that (II.1.1) and (1.3) define distinct additive spin systems with these flip rates. Note that $\emptyset$ is a trap for the canonical system $\{(\xi_t^A)\}$, but not for $\{(\bar{\bar{\xi}}_t^A)\}$.

(1.5) Problems. Prove that $\{(\bar{\xi}_t^A)\}$ and $\{(\bar{\bar{\xi}}_t^A)\}$, as defined by (1.2) and (1.3) respectively, are always Feller systems. Conclude that in the case of uniqueness, the canonical system $\{(\xi_t^A)\}$ given by (II.1.1) is always Feller. Observe that when weak influence from ∞ arises, $\{(\bar{\xi}_t^A)\}$ and $\{(\bar{\bar{\xi}}_t^A)\}$ are distinct.

(1.6) Problem. Give an example where $\{(\bar{\xi}_t^A)\}$, as defined by (1.2), does not coincide with the canonical $\{(\xi_t^A)\}$. Is $\{(\xi_t^A)\}$ Feller in this case?

Uniqueness, like ergodicity, can be studied with the aid of dual processes. Say that there is an explosion from $(x_0, 0)$ at time t in $\hat{P}$ if there are sites

$x_0, x_1, \cdots$ in Z^d and times $0 = t_0 < t_1 \le t_2 \le \cdots$ such that

(i) there is a path from (x_0, t_0) to (x_n, t_n) for each $n \ge 1$,

(ii) $\lim_{n\to\infty} |x_n| = \infty$,

(iii) $\lim_{n\to\infty} t_n = t < \infty$.

Condition (1.1) is clearly equivalent to

$$\hat{P}\text{ (no explosions)} = 1 . \tag{1.7}$$

Our immediate goal is to prove uniqueness of the additive and cancellative systems induced by $\mathcal{P}$ when (1.1) holds. We will thereby obtain uniqueness for all of the special systems discussed in Chapters 2 and 3.

Given <u>any</u> percolation substructure $\mathcal{P}$, one can define the <u>minimal dual processes</u> $(\check{\xi}_t^B)$ and $(\check{\eta}_t^B)$, $B \in S_0$, in terms of the dual substructure $\hat{\mathcal{P}}$. To do so, introduce the first explosion time $\check{\tau}^B$,

$$\check{\tau}^B = \inf\{t : \exists \text{ explosion at } t \text{ from some } (x, 0),\ x \in B\} \tag{1.8}$$

($= \infty$ if there is no explosion from $(B, 0)$). Define the minimal dual processes $(\check{\xi}_t^B)$ and $(\check{\eta}_t^B)$ precisely as in Chapters 2 and 3 for times $t < \check{\tau}^B$, and send them to the trap Δ at time $\check{\tau}^B$. Let $\check{G}$ be the "Q-matrix" operator for the minimal dual system of denumerable Markov processes; $\check{G}f$ is defined for bounded functions f.

Finally, introduce the functions

$$\begin{aligned} f_B(A) &= 1 \quad \text{if} \quad A \cap B = \emptyset \\ &= 0 \quad \text{if} \quad A \cap B \neq \emptyset \\ f_B(\Delta) &\equiv 0 , \end{aligned}$$

$A, B \in S$, and

$$\begin{aligned} g_B(A, 0) &= 1 \text{ if } |A \cap B| \text{ even} \\ &= 0 \text{ if } |A \cap B| \text{ odd} \\ g_B(A, 1) &= 1 - g_B(A, 0) , \quad g_B(\Delta) \equiv 0 , \end{aligned}$$

$A, B \in S$ and at least one finite. Also, put

$$g_B(A) = g_B(A, 0) \qquad A \in S, \quad B \in S_0 .$$

In the next section we will use minimal dual processes and the "function bases" $\{f_B(A)\}$ and $\{g_B(A, \varepsilon)\}$ to obtain uniqueness theorems for additive and cancellative particle systems.

(1.9) Notes. The material in this chapter extends results of Gray and Griffeath (1977), but the graphical approach is new. Some other papers on the uniqueness of particle systems which apply in greater generality are Liggett (1972), Sullivan (1974), Higuchi and Shiga (1975), Holley and Stroock (1976a), Gray and Griffeath (1976) and Liggett (1977).

2. Uniqueness theorems.

We now state and prove two lemmas which connect systems with pregenerator G to the minimal dual with Q-matrix $\check{G}$.

(2.1) Lemma. If $\{(\xi_t^A)\}$ is additive with pregenerator G, and $\check{G}$ is the corresponding dual operator,

$$G f_B(A) = \check{G} f_A(B) \qquad A \in S, \quad B \in S_0 .$$

Similarly, if $\{(\eta_t^A)\}$ is cancellative, with pregenerator G and dual operator $\check{G}$, then

$$G g_B(A) = \check{G} g_A(B, 0) \qquad A \in S, \quad B \in S_0 .$$

Proof. We check only the additive case, since the cancellative computation is very similar. The additive dual operator $\check{G}$ obtained from $\hat{\mathcal{P}}$ has the form

$$\check{G} f(B) = \sum_{i,x} \lambda_{i,x} [f(\check{\mathfrak{A}}_{i,x}(B)) - f(B)] ,$$

where

$$\check{\mathfrak{U}}_{i,x}(B) = \bigcup_{y\in B} \hat{W}_{i,x}(y) \qquad \text{if} \quad V_{i,x} \cap B = \emptyset$$

$$= \Delta \qquad \text{if} \quad V_{i,x} \cap B \neq \emptyset \ .$$

Fix $A \in S$, $B \in S_0$, and note that

$$\mathfrak{U}_{i,x}(A) \cap B = \emptyset \quad \text{if and only if} \quad \check{\mathfrak{U}}_{i,x}(B) \cap A = \emptyset \ \underline{\text{and}} \ V_{i,x} = \emptyset \ .$$

Thus,

$$Gf_B(A) = \sum_{i,x} \lambda_{i,x} [f_B(\mathfrak{U}_{i,x}(A)) - f_B(A)]$$

$$= \sum_{\substack{i,x: \\ V_{i,x} \cap B = \emptyset}} \lambda_{i,x} [f_A(\check{\mathfrak{U}}_{i,x}(B)) - f_A(B)]$$

$$+ \sum_{\substack{i,x: \\ V_{i,x} \cap B \neq \emptyset}} \lambda_{i,x} [f_A(\Delta) - f_A(B)]$$

$$= \check{G} f_A(B) \ . \qquad \square$$

For the lemma which follows, and the rest of the chapter, we introduce the notation

$$P^t f(A) = E[f(\xi_t^A)] \ ,$$

$$u_A^t(B) = P^t f_B(A) \ ,$$

$$\check{P}^t f(B) = \hat{E}[f(\check{\xi}_t^B)] \ ,$$

$A \in S$, $B \in S_0$. Note that $P^t f_B(A) = P(\xi_t^A \cap B = \emptyset)$ are simply the basic cylinder probabilities of the additive theory. Similarly $P^t g_B(A) = 2P(|\eta_t^A \cap B| \text{ even}) - 1$ captures the basic cancellative probabilities.

(2.2) Lemma. Let $\{(\xi_t^A)\}$ be additive, with pregenerator G and dual operator $\check{G}$. For all $t \geq 0$, $A \in S$, $B \in S_0$,

$$\frac{du_A^t(B)}{dt} = \check{G} u_A^t(B) \ ,$$

and

$$u_A^0(B) = f_A(B) .$$

Analogous equations hold in the cancellative setting, with $f_B(A)$ and $f_A(B)$ replaced by $g_B(A)$ and $g_A(B,0)$ respectively.

Proof. We present only the additive case. Fix $A \in S$, $B \in S_0$, and write $\check{\gamma}_B(A) = \check{G} f_A(B)$. Then by standard semigroup theory, Lemma (2.1) and Fubini's theorem,

$$\begin{aligned}\frac{du_A^t(B)}{dt} &= \frac{dP^t f_B(A)}{dt} \\ &= P^t G f_B(A) \\ &= P^t \check{\gamma}_B(A) \\ &= \check{G} u_A^t(B) .\end{aligned}$$

The equation for $t = 0$ is trivial. □

Thus, $u_A^t(B)$ and $\check{P}^t f_A(B)$ satisfy the same "backward" differential equation with the same boundary condition at $t = 0$. A general uniqueness theorem follows from this fact. To state it, we introduce

$$\check{\tau}_n^B = \inf \{ t < \check{\tau}_\Delta^B : \check{\xi}_t^B \cap b_n(0)^c_- \neq \emptyset \}$$

(the empty inf is $+\infty$) in the additive case, with the analogous definition in the cancellative case.

(2.3) Theorem. If for every additive system $\{(\xi_t^A)\}$ induced by a given substructure $\mathcal{P}$ (i.e. for every system with the additive pregenerator G induced by $\mathcal{P}$),

$$\text{(2.4)} \qquad \lim_{n\to\infty} \hat{E}[u_A^{t-\check{\tau}_n^B}(\check{\xi}^B_{\check{\tau}_n^B}), \check{\tau}_n^B \le t] = 0 \qquad \forall A \in S, \quad B \in S_0 ,$$

then there is a unique additive system induced by $\mathcal{P}$. It is Feller, and may be defined by (II.1.1). An analogous assertion holds for cancellative systems; when unique, $\{(\eta_t^A)\}$ is Feller, and may be defined by (III.1.1).

Before proceeding to prove Theorem (2.3), we derive the important fact that (1.7) (or equivalently (1.1)) yields uniqueness.

(2.5) Corollary. If $\hat{P}(\text{no explosions}) = 1$, then there is a unique additive system and a unique cancellative system induced by ρ.

Proof. The expectation in (2.4) and its analogue in the cancellative setting are majorized in absolute value by $\hat{P}(\check{\tau}_n^B \le t)$, and

$$\lim_{n\to\infty} \hat{P}(\check{\tau}_n^B \le t) = \hat{P}\,(\text{explosion from } (B,0) \text{ by time } t) = 0 . \qquad \square$$

Proof of Theorem (2.3). Fix $A \in S$, $B \in S_0$, $t > 0$, and write $\check{\tau}_n = \check{\tau}_n^B \wedge t$, $\check{\xi}_r = \check{\xi}_r^B$. For any function $f(t-r, \check{\xi}_r)$, $0 \le r \le t$, if we set

$$M_s = f((t-s)^+, \check{\xi}(s \wedge t)) - \int_0^{s \wedge t} (\tfrac{d}{dr} + \check{G}) f(t-r, \check{\xi}_r)\, dr ,$$

then $M_{s \wedge \check{\tau}_n}$ is a $\hat{P}$-martingale for each n. Take $f(t-r, \check{\xi}_r) = u_A^{t-r}(\check{\xi}_r)$, and note that for $r \in [0, \check{\tau}_n]$,

$$(\tfrac{d}{dr} + \check{G}) f(t-r, \check{\xi}_r) = (\tfrac{d}{dr} + \check{G}) u_A^{t-r}(\check{\xi}_r) = 0$$

by Lemma (2.2). Thus the equation $M_0 = \hat{E}[M_{\check{\tau}_n}]$ becomes

$$u_A^t(B) = \hat{E}[f_A(\check{\xi}_t), \check{\tau}_n > t] + \hat{E}[u_A^{t-\check{\tau}_n}(\check{\xi}_{\check{\tau}_n}), \check{\tau}_n \le t] .$$

If (2.4) holds, then letting $n \to \infty$ we get

(2.6) $$u_A^t(B) = \hat{P}[\check{\xi}_t^B \cap A = \emptyset,\ \check{\tau}_n > t \text{ eventually in } n] .$$

Thus μP^t is uniquely defined for all $\mu \in \mathfrak{M}$, $t \ge 0$, and (II.1.1) must be an appropriate representation. To see that $\{(\xi_t^A)\}$ is Feller, either use Problem (1.5) or derive the Feller property directly from (2.6). The proof in the cancellative setting is virtually identical. $\square$

It turns out that uniqueness can arise even when explosions occur. We illustrate this possibility with a uniqueness result for proximity systems. By analogy to the definitions preceding (II.1.4), say that the minimal dual process

$(\check{\xi}_t^B)$ has a <u>weak explosion at</u> $\check{\tau}^B$ if

$$|\check{\xi}^B_{\check{\tau}^B_-}| < \infty ,$$

and a <u>strong explosion at</u> $\check{\tau}^B$ if

$$|\check{\xi}^B_{\check{\tau}^B_-}| = \infty ,$$

where $\check{\xi}_{\check{\tau}^B_-} = \lim_{t \uparrow \check{\tau}^B} \check{\xi}_t^B$. Roughly speaking, if all explosions are strong and spontaneous births occur at a positive minimal rate, then uniqueness holds.

(2.7) <u>Theorem</u>. Let $\{(\xi_t^A)\}$ be a (canonical) extralineal proximity process, with flip rates of the form (II.2.12), $\wp$ its underlying substructure. If

(i) $\inf_x \kappa_x = \kappa > 0$,

(ii) $\sup_{\substack{x, A: \\ x \in A}} c_x(A) = K < \infty$,

and

(iii) $\check{P}$ (weak explosion at $\check{\tau}^B$) = 0 $\qquad \forall B \in S_0$,

then there is a unique system $\{(\xi_t^A)\}$ with pregenerator G induced by $\wp$.

<u>Proof:</u> It suffices to check (2.4). An appropriate decomposition yields

$$\hat{E}[u_A^{t-\check{\tau}_n^B}(\check{\xi}^B_{\check{\tau}_n^B}), \check{\tau}_n^B \le t]$$

(a_n) $\qquad \le \hat{P}(\check{\tau}_n^B \in [t-\delta, t])$

(b_n) $\qquad + \hat{P}(\check{\tau}_n^B < t-\delta, |\check{\xi}_{\check{\tau}_n^B}^B| \ge M) \cdot \sup_{B: |B| \ge M} \sup_{s \ge \delta} u_A^s(B)$

(c_n) $\qquad + \hat{P}(\check{\tau}_n^B < t, |\check{\xi}^B_{\check{\tau}_n^B}| < M)$

for arbitrary $\delta \in (0, t]$, $M \ge 0$. Now

$\limsup_{n\to\infty} a_n = \hat{P}(\check{\tau}_n^B \in [t-\delta, t]$ for infinitely many $n) = \hat{P}(\check{\tau}^B \in [t-\delta, t])$.
Let $\check{\tau}_*^B$ be the time of the first jump by $(\check{\xi}_t^B)$. Since $\check{\tau}^B = \check{\tau}_*^B + (\check{\tau}^B - \check{\tau}_*^B)$ is the independent sum of an exponential variable and the remaining time, the distribution of $\check{\tau}^B$ is absolutely continuous on $(0, \infty)$. Thus we can choose $\delta > 0$ so that $\limsup_{n\to\infty} a_n \leq \frac{\varepsilon}{2}$. Next we take M large enough that $b_n \leq \frac{\varepsilon}{2}$ for all n. To see that this is possible, we first estimate

$$\begin{aligned}\frac{du_A^s(B)}{ds} &= E[Gf_B(\xi_s^A)] \\ &= \sum_{x\in B} E[-c_x(\xi_s^A),\ \xi_s^A \cap B = \emptyset] \\ &\quad + \sum_{x\in B} E[c_x(\xi_s^A),\ \xi_s^A \cap (B-x) = \emptyset,\ \ x \in \xi_s^A] \\ &\leq -|B|\,\kappa u_A^s(B) + K \sum_{x\in B} P(\xi_s^A \cap (B-x) = \emptyset,\ \ x \in \xi_s^A) \\ &\leq -|B|\,\kappa u_A^s(B) + K[1 - u_A^s(B)]\ .\end{aligned}$$

By Gronwall's inequality,

$$u_A^s(B) \leq \frac{K}{|B|\,\kappa + K} + \frac{|B|\,\kappa}{|B|\,\kappa + K}\, e^{-(|B|\,\kappa + K)s}$$

which tends to 0 uniformly in $s \geq \delta$ as $|B| \to \infty$. This controls the second term of b_n; bound the first by 1 to establish the chain. Finally, with δ and M so chosen, note that $\lim_{n\to\infty} c_n = \hat{P}(\check{\tau}^B \leq t,\ |\check{\xi}_{\check{\tau}^B-}^B| \leq M) = 0$, again by hypothesis. Thus (2.4) holds, and uniqueness is proved. □

(2.8) <u>Example</u>. Consider the extralineal proximity system on Z with flip rates of the form

$$\begin{aligned} c_x(A) &= 0 \quad \text{if} \quad x < 0 \quad \text{or} \quad x \in A \\ &= 1 \quad \text{if} \quad 0 > x \notin A \ \ \underline{\text{or}}\ \ x \geq 0 \ \text{ and } \ A \cap [0, x+1] = \emptyset \\ &= 100^{x^2} \quad \text{otherwise.}\end{aligned}$$

The substructure $\mathcal{P}$ for this system has β's at each site x at rate 1, and for $x \geq 0$, arrows arrive at x from every site in $[0, x+1]$ at rate 100^{x^2}. It is easy to check the hypotheses of Theorem (2.7), so uniqueness holds. This example exhibits an unusual phenomenon in the theory of Markovian semigroups, which merits a brief discussion. Given pregenerator G, one defines the closure $\overline{G}$ of G by

$$\text{graph } (\overline{G}) = \overline{\text{graph } (G)} \qquad (\text{in } C \times C).$$

Thus if $\overline{G}$ has domain $\mathcal{D}(\overline{G})$ and $h \in \mathcal{D}(\overline{G})$, then there are functions $f_n \in \mathfrak{F}$ such that $\|h - f_n\| \to 0$ and $\|\overline{G}h - Gf_n\| \to 0$ as $n \to \infty$. As a rule, if there is uniqueness for G, then $\overline{G}$ is the generator of $\{(\xi_t^A)\}$. In particular this is the case whenever the Hille-Yosida Theorem applies. For the present example, however, if $G^e : \mathcal{D}(G^e) \to C$ is the generator of $\{(\xi_t^A)\}$, one can find a function $\varphi \in \mathcal{D}(G^e)$ such that whenever $f \in \mathfrak{F}$ and $\|\varphi - f\| \leq \frac{1}{100}$, then $\|G^e\varphi - Gf\| > \frac{1}{100}$, i.e. $\varphi \in \mathcal{D}(G^e) - \mathcal{D}(\overline{G})$. Thus the unique generator extending G is not $\overline{G}$ in this case.

(2.9) Notes. Theorem (2.3) and Corollary (2.5) are adaptations to the additive setting of results from Holley, Stroock and Williams (1977). Theorem (2.7) and Example (2.8) are taken from Gray and Griffeath (1977).

3. Nonuniqueness examples.

In this final section we briefly discuss nonuniqueness possibilities for particle systems. To keep matters simple, only the additive setting will be considered. One of the simplest nonuniqueness examples was encountered already in Problem (1.4). For those flip rates, the presence of weak influence from ∞ in $\mathcal{P}$ gives rise to distinct systems defined by (II.1.1) and (1.3). In fact, there is a continuum of systems with the flip rates of (1.4). Indeed, any substructure $\mathcal{P}$ with weak influence from ∞ gives rise to an infinite family of Feller additive systems.

(3.1) Theorem. Let G be the additive pregenerator induced by a percolation substructure $\mathcal{P}$. If

$$\hat{P}(\check{\tau}^B < \infty \,,\ |\xi^{\check{B}}_{\check{\tau}^B_-}| < \infty) > 0 \quad \text{for some} \quad B \in S_0 \,,$$

then there is a continuum of distinct Feller additive systems with pregenerator G. More precisely, let Λ be the maximal set in Z^d such that $\Lambda \subset \xi^{\check{B}}_{\check{\tau}^B_-}$ $\hat{P}$ - a.s. Then to each probability measure π on $S_0 \cup \{\Delta\}$, there corresponds a Feller system $\{(\xi^A_{\pi,t})\}$ with pregenerator G and semigroup (P^t_π), where $(P^t_\pi) \neq (P^t_{\pi'})$ if $\pi|_{Z^d-\Lambda} \neq \pi'|_{Z^d-\Lambda}$.

<u>Sketch of proof</u>: Given a probability measure π on $S_0 \cup \{\Delta\}$, let Λ_t, $t \in (0,\infty)$ be π-distributed independent random variables. Adjoin an isolated point ∞ to Z^d and extend $\mathcal{P}$ to a percolation substructure $\mathcal{P}_\infty$ on $(Z^d \cup \{\infty\}) \times T$ as follows. If $\Lambda_t = \Delta$, label (∞, t) with a β, while if $\Lambda_t \in S_0$, draw arrows from (Λ_t, t) to (∞, t). We may think of these arrows as allowing for "influence through ∞." Now say that

$x \in \xi^A_{\pi,t}$ if

(i) there is a path up to (x,t) from $(A,0)$, possibly "through ∞,"

<u>or</u>

(ii) there is a path up to (x,t) from some (y,s) labelled β, $y \in Z^d \cup \{\infty\}$, the path again possibly "through ∞,"

<u>or</u>

(iii) there is strong influence from ∞ to (x,t)

<u>or</u>

(iv) there is a path down from (x,t) in the reverse substructure whose visits to ∞ have an accumulation point.

A path only enters $Z^d \times T$ from (∞, s) if the reverse path "wanders off to ∞" at time s. We leave the precise formulation of the effects involving ∞, as well as the details of the construction, to the interested reader. One can check that the system $\{(\xi^A_{\pi,t})\}$ so defined is Feller with pregenerator G, and that under the hypotheses of the theorem, different π's give rise to different systems. □

(3.2) Problems. Which measure π gives rise to the system $\{(\bar{\xi}^A_t)\}$ defined by (1.2)? Which π yields the system $\{(\bar{\bar{\xi}}^A_t)\}$ of (1.3)? For the flip rates of Problem (1.4), construct a translation invariant system such that both $\emptyset$ and Z are traps, and one such that neither $\emptyset$ nor Z is a trap. Under the hypotheses of Theorem (3.1), find additional nonuniqueness examples which are not covered by the construction sketched above.

(3.3) Problem. Let $\{(\xi^A_t)\}$ be a spin system on Z with flip rates

$$
\begin{aligned}
c_x(A) &= 0 && x \le -1, \\
&= r_0[A(0) + (1-2A(0))A(1)] && x = 0, \\
&= r_x[A(x) + (1-2A(x))(p_x A(x+1) + q_x A(x-1))] && x \ge 1,
\end{aligned}
$$

for some $r_x > 0$, $0 < p_x < 1$, with $q_x = 1 - p_x$. Show that $\{(\xi^A_t)\}$ is additive. Describe the dual processes $(\check{\xi}^A_t)$, in particular the one-particle duals $(\check{\xi}^x_t)$, $x \in Z$. For general r's and p's, discuss as many kinds of nonuniqueness examples as you can find. (You may want to make use of Chapter IV of Dynkin and Yushkevich (1969).)

(3.4) Notes. The material of this section is based on Gray and Griffeath (1977), although the graphical approach is new. The simple nonuniqueness example of (1.4) first appeared in Gray and Griffeath (1976). Another nonuniqueness example may be found in Holley and Stroock (1976a).

Bibliography

Omer Adelman (1976). Some use of some "symmetries" of some random processes, Ann. Inst. Henri Poincaré 12, 193-197.

F. Bertein and A. Galves (1978). Une classe de systèmes de particules stable par association, Z. Wahrscheinlichkeitstheorie Verw. Geb. 41, 73-85.

D. Blackwell (1958). Another countable Markov process with only instantaneous states, Annals of Mathematical Statistics 29, 313-316.

M. Bramson and D. Griffeath (1978a). Renormalizing the 3-dimensional voter model, Annals of Probability, to appear.

M. Bramson and D. Griffeath (1978b). Clustering and dispersion rates for some interacting particle systems on Z, Annals of Probability, to appear.

S. Broadbent and J. Hammersley (1957). Percolation processes I. Crystals and mazes, Proc. Cambridge Phil. Soc. 53, 629-645.

J. Chover (1975). Convergence of a local lattice process, Stochastic Processes and their Applications 3, 115-135.

P. Clifford and A. Sudbury (1973). A model for spatial conflict, Biometrika 60, 581-588.

D. Dawson (1974a). Information flow in discrete Markov systems, Journal of Applied Probability 11, 594-600.

D. A. Dawson (1974b). Discrete Markov Systems, Carleton Math. Lecture Notes no. 10, Carleton University, Ottawa.

D. Dawson (1975). Synchronous and asynchronous reversible Markov systems. Canadian Mathematical Bulletin 17, 633-649.

D. A. Dawson (1978). The critical measure diffusion process, Z. Wahrscheinlichkeitstheorie Verw. Geb. 40, 125-145.

D. Dawson and G. Ivanoff (1978). Branching diffusions and random measures, Advances in Probability 5, ed. A. Joffe, P. Ney, Dekker, New York.

R. L. Dobrushin (1971). Markovian processes with a large number of locally interacting components, Problems of Information Transmission 7, 149-164 and 235-241.

R. Durrett (1978). An infinite particle system with additive interactions, to appear.

E. B. Dynkin and A. A. Yushkevich (1969). Markov Processes. Theorems and Problems, Plenum Press, New York.

P. Erdös and P. Ney (1974). Some problems on random intervals and annihilating particles, Annals of Probability 2, 828-839.

J. Fleischmann (1978). Limit theorems for critical branching random fields, T.A.M.S. 239, 353-389.

H. O. Georgii (1976). Stochastische Prozesse Für Interaktionssysteme. Heidelberg.

R. Glauber (1963). The statistics of the stochastic Ising model, Journal of Mathematical Physics 4, 294-307.

L. Gray (1978). Controlled spin systems, Annals of Probability 6, 953-974.

L. Gray and D. Griffeath (1976). On the uniqueness of certain interacting particle systems, Z. Wahrscheinlichkeitstheorie verw. Geb. 35, 75-86.

L. Gray and D. Griffeath (1977). On the uniqueness and nonuniqueness of proximity processes, Annals of Probability 5, 678-692.

D. Griffeath (1975). Ergodic theorems for graph interactions, Advances in Applied Probability 7, 179-194.

D. Griffeath (1977). An ergodic theorem for a class of spin systems, Ann. Inst. Henri Poincaré 13, 141-157.

D. Griffeath (1978a). Limit theorems for nonergodic set-valued Markov processes. Annals of Probability 8, 379-387.

D. Griffeath (1978b). Annihilating and coalescing random walks on Z_d. Z. Wahrscheinlichkeitstheorie verw. Geb. 46, 55-65.

D. Griffeath (1979). Pointwise ergodicity of the basic contact process, Annals of Probability 7, 139-143.

R. Griffiths (1972). The Peierls argument for the existence of phase transitions, Mathematical Aspects of Statistical Mechanics, J.C.T. Pool (ed.), SIAM-AMS Proceedings, Providence, Amer. Math. Soc. 5, 13-26.

J. Hammersley (1959). Bornes supérieures de la probabilité critique dans un processus de filtration. Le Calcul des Probabilités et ses Applications. Centre National de la Recherche Scientifique, Paris, 17-37.

T. E. Harris (1960). Lower bound for the critical probability in a certain percolation process, Proc. Cambridge Phil. Soc. 56, 13-20.

T. E. Harris (1972). Nearest neighbor Markov interaction processes on multi-dimensional lattices, Advances in Mathematics 9, 66-89.

T. E. Harris (1974). Contact interactions on a lattice, Annals of Probability 2, 969-988.

T. E. Harris (1976). On a class of set-valued Markov processes, Annals of Probability 4, 175-194.

T. E. Harris (1977). A correlation inequality for Markov processes in partially ordered state spaces, Annals of Probability 5, 451-454.

T. E. Harris (1978). Additive set-valued Markov processes and percolation methods, Annals of Probability 6, 355-378.

L. L. Helms (1974). Ergodic properties of several interacting Poisson particles, Advances in Mathematics 12, 32-57.

Y. Higuchi and T. Shiga (1975). Some results on Markov processes of infinite lattice spin systems, Journal of Mathematics of Kyoto University 15, 211-229.

R. Holley (1970). A class of interactions in an infinite particle system, Advances in Mathematics 5, 291-309.

R. Holley (1971). Free energy in a Markovian model of a lattice spin system, Communications in Mathematical Physics 23, 87-99.

R. Holley (1972a). Markovian interaction processes with finite range interactions, Annals of Mathematical Statistics 43, 1961-1967.

R. Holley (1972b). An ergodic theorem for interacting systems with attractive interactions, Z. Wahrscheinlichkeitstheorie verw. Geb. 24, 325-334.

R. Holley (1974). Recent results on the stochastic Ising model, Rocky Mountain Journal of Mathematics 4, 479-496.

R. Holley and T. M. Liggett (1975). Ergodic theorems for weakly interacting infinite systems and the voter model, Annals of Probability 3, 643-663.

R. Holley and T. M. Liggett (1978). The survival of contact processes, Annals of Probability 6, 198-206.

R. Holley and D. Stroock (1976a). A martingale approach to infinite systems of interacting processes, Annals of Probability 4, 195-228.

R. Holley and D. Stroock (1976b). Applications of the stochastic Ising model to the Gibbs states, Communications in Mathematical Physics 48, 249-266.

R. Holley and D. Stroock (1976c). L_2 theory for the stochastic Ising model, Z. Wahrscheinlichkeitstheorie verw. Geb. 35, 87-101.

R. Holley and D. Stroock (1976d). Dual processes and their application to infinite interacting systems, Advances in Mathematics, to appear.

R. Holley and D. Stroock (1978). Nearest neighbor birth and death processes on the real line, Acta Math. 140, 103-154.

R. Holley, D. Stroock and D. Williams (1977). Applications of dual processes to diffusion theory, Proc. Sympos. Pure Math. 31, 23-36. Amer. Math. Soc., Providence, R.I.

F. P. Kelly (1977). The asymptotic behavior of an invasion process, Journal of Applied Probability 14, 584-590.

W. C. Lee (1974). Random stirring of the real line, Annals of Probability 2, 580-592.

T. M. Liggett (1972). Existence theorems for infinite particle systems, T.A.M.S. 165, 471-481.

T. M. Liggett (1973). A characterization of the invariant measures for an infinite particle system with interactions, T.A.M.S. 179, 433-453.

T. M. Liggett (1974). A characterization of the invariant measures for an infinite particle system with interactions II, T.A.M.S. 198, 201-213.

T. M. Liggett (1975). Ergodic theorems for the asymmetric simple exclusion process, T.A.M.S. 213, 237-261.

T. M. Liggett (1976). Coupling the simple exclusion process, Annals of Probability 4, 339-356.

T. M. Liggett (1977). The stochastic evolution of infinite systems of interacting particles, Lecture Notes in Mathematics 598, 187-248. Springer-Verlag, New York.

T. M. Liggett (1978). Attractive nearest neighbor spin systems on the integers, Annals of Probability 6, 629-636.

K. Logan (1974). Time reversible evolutions in statistical mechanics, Cornell University, Ph.D. dissertation.

J. C. Lootgieter (1977). Problèmes de récurrence concernant des mouvements aléatoires de particules sur Z avec destruction, Ann. Inst. Henri Poincaré 13, 127-139.

V. A. Malyšev (1975). The central limit theorem for Gibbsian random fields, Soviet Math. Dokl. 16, 1141-1145.

N. Matloff (1977). Ergodicity conditions for a dissonant voter model, Annals of Probability 5, 371-386.

D. Mollison (1977). Spatial contact models for ecological and epidemic spread, J. Royal Statistical Soc. B, 39, 283-326.

C. J. Preston (1974). Gibbs states on countable sets, Cambridge University Press.

D. Richardson (1973). Random growth in a tesselation, Proc. Cambridge Phil. Soc. 74, 515-528.

S. Sawyer (1976). Results for the stepping stone model for migration in population genetics, Annals of Probability 4, 699-728.

S. Sawyer (1978). A limit theorem for patch sizes in a selectively-neutral migration model, to appear.

D. Schwartz (1976). Ergodic theorems for an infinite particle system with births and deaths, Annals of Probability 4, 783-801.

D. Schwartz (1977). Applications of duality to a class of Markov processes, Annals of Probability 5, 522-532.

D. Schwartz (1978). On hitting probabilities for an annihilating particle model, Annals of Probability 6, 398-403.

V. K. Shante and S. Kirkpatrick (1971). Introduction to percolation theory, Advances in Physics 20, 325-357.

F. Spitzer (1970). Interaction of Markov processes, Advances in Mathematics 5, 246-290.

F. Spitzer (1971). Random fields and interacting particle systems, M.A.A. Summer Seminar Notes, Williamstown, Mass.

F. Spitzer (1974a). Recurrent random walk of an infinite particle system, T.A.M.S. 198, 191-199.

F. Spitzer (1974b). Introduction aux processus de Markov à paramètres dans Z_ν, Lecture Notes in Mathematics 390, Springer-Verlag, New York.

F. Spitzer (1976). Principles of Random Walk, 2nd ed., Springer-Verlag, New York.

F. Spitzer (1977). Stochastic time evolution of one dimensional infinite particle systems, B.A.M.S. 83, 880-890.

O. N. Stavskaya (1975). Sufficient conditions for the uniqueness of a probability field and estimates for correlations, Matematicheskie Zametki 18, 609-620.

O. N. Stavskaya and I. I. Pyatetskii-Shapiro (1968). On homogeneous networks of spontaneously active elements, Problemy kibernetiki, Nauka Moscow 20, 91-106.

D. Stroock (1978). Lectures on Infinite Interacting Systems, Kyoto University Lectures in Math. no. 11.

W. G. Sullivan (1974). A unified existence and ergodic theorem for Markov evolution of random fields, Z. Wahrscheinlichkeitstheorie verw. Geb. 31, 47-56.

W. G. Sullivan (1975). Markov processes for random fields, Communications Dublin Inst., Advanced Studies A, No. 23.

W. G. Sullivan (1976). Processes with infinitely many jumping particles, P.A.M.S. 54, 326-330.

A. L. Toom (1968). A family of uniform nets of formal systems, Soviet Mathematics 9, 1338-1341.

A. L. Toom (1974). Nonergodic multidimensional systems of automata, Problemy Peredachi Informatsii 10, 239-246.

L. N. Vasershtein (1969). Markov processes over denumerable products of spaces, describing large systems of automata, Problemy Peredachi Informatsii 5 (3), 64-73.

L. N. Vasershtein and A. M. Leontovich (1970). Invariant measures of certain Markov operators describing a homogeneous random medium, Problemy Peredachi Informatsii 6 (1), 71-80.

N. B. Vasilev (1969). Limit behavior of one random medium, Problemy Peredachi Informatsii 5 (4), 68-74.

N. B. Vasil'ev, L. G. Mityushin, I. I. Pyatetskii-Shapiro and A. L. Toom (1973). Stavskaya Operators (in Russian), Preprint no. 12, Institute of Applied Math., Academy of Sciences, U.S.S.R., Moscow.

T. Williams and R. Bjerknes (1972). Stochastic model for abnormal clone spread through epithelial basal layer, Nature 236, 19-21.

Subject Index